THE
PEAT GARDEN
AND
ITS PLANTS

Alfred Evans

THE
PEAT GARDEN
AND
ITS PLANTS

with 16 pages of colour photographs,
71 monochrome photographs, and drawings in text

J M Dent & Sons Limited London
in association with The Royal Horticultural Society

First published 1974

© Alfred Evans, 1974

All rights reserved. No part of this
publication may be reproduced, stored
in a retrieval system, or transmitted,
in any form or by any means, electronic,
mechanical, photocopying, recording or
otherwise, without the prior permission
of J. M. Dent & Sons Ltd.
Made in Great Britain
at the
Aldine Press · Letchworth · Herts
for
J. M. DENT & SONS LTD
Aldine House · Albemarle Street · London

This book is set in 11 on 12 point Centaur 252

ISBN 0 460 04113 4

CONTENTS

LIST OF PLATES

COLOUR

MONOCHROME

BETWEEN PAGES 36 AND 37

BETWEEN PAGES 116 AND 117

TEXT FIGURES

FOREWORD by G. S. Thomas, V.M.H.

One of the richest collections of plants in the British Isles is in the Royal Botanic Garden, Edinburgh. Of all the areas in it, the one that fascinates me most is that containing the heath garden, the rock garden and the peat beds in sheltered places among trees. Alfred Evans has been in charge of these areas for many years and I doubt if a better guide could be found for a tour of all the delectable plants which are grown. He has been engaged in peat gardening since its inception—in fact he is acknowledged the leading expert, and what he has spread before us in this book is enough to whet the appetite of any gardener, whether experienced or not. Peat gardening has come to stay and while in the cooler north plants thrive in it beyond belief, it is also a godsend to those who garden in the drier parts of southern England. Further, it has the added advantage of being a good method of cultivation for the tiniest gardens as well as those on a larger scale. I think there is no doubt that his book will bring this fact home to those who as yet have not realized how difficult plants can be made successful in peat, while the easy ones thrive as never before.

ACKNOWLEDGMENTS

It would be quite difficult if not impossible to write a book without relying on others for help, especially a book of this nature. I did not set out to try to do this and so there are a number of kind people to whom I must give credit now that the work is done.

In the first instance I am grateful to the Department of Agriculture and Fisheries for Scotland for granting permission for the book to be written and to Mr D. M. Henderson, Regius Keeper of the Royal Botanic Garden, Edinburgh, who, recognizing the need for such a publication, encouraged me and allowed me the use of the Garden's excellent facilities. Readers and checkers, too, must be identified and I am grateful to Miss Dorothea Purves, the Plant Records Officer, for the painstaking way in which she scanned the manuscript and checked names, the last-mentioned task being one in a subject which appears to many to be highly unstable at best.

Constructive criticism is of immense value and I appreciate the thorough way in which this was provided by Mr Graham Thomas, the Publisher's Horticultural Adviser. To Miss Rosemary Smith I acknowledge the skill used in supplying line drawings while Mr Ross Eudall used his technical know-how in producing black and white enlargements from my 35mm negatives.

To all those mentioned I am indebted but surely this is also the time to recognize those horticulturists who have gone before and who recorded their interpretations of plant responses and so added to our knowledge. To the intrepid plant collectors who have made our gardens rich and to the truth-searching botanist, however much he may be maligned, who bravely attempts to bring order and identity to the world's flora, much is owed.

A. Evans

TO BETH

not solely for her patience and tolerance
but also for her constant encouragement

INTRODUCTION

The task of writing a book on any subject is not something which can be undertaken lightly and, although I have been associated with peat gardens for almost twenty-five years, I thought quite seriously about tackling such a project. After all, so much has been written about aspects and plants that I wondered if there were anything new about which I could write. The Peat Garden, certainly, is a plantsman's garden and as such can only be maintained by someone who is deeply interested in plants and who is prepared to tend them with only the minimum of tools.

I am aware that modern techniques are being developed all the time; glasshouse structures are being made more standard and automation in the control of heating, ventilation and irrigation is here to stay. Selective weedkillers and systemic insecticides, even fungicides, are on the market, all with a view to making life easier for the gardener. No doubt these advancements in building methods and discoveries in new chemicals and their application do make life easier but—and here I must be serious and keep in view the standard of the end product—can it be claimed that these new aids have made it possible to produce better plants than were ever raised before? I doubt it. However, if it can be said that just as many good plants as have ever been grown can be raised today with the addition of this equipment, and that these new techniques have helped to remove the backache, the humdrum and tedious tasks in gardening, then they are worth considering. On the other hand, if it has to be admitted that these modifications have in fact reduced the standard of cultivation and in some cases have even brought a level of mediocrity, then they have failed so far as the enthusiast is concerned. In some instances this slight drop in standards, where the machine takes over, may have to be accepted, albeit reluctantly. But where this argument is used on the enthusiastic plantsman he can only stand, stare, listen and no doubt be unable to comprehend anyone having an attitude which could be satisfied with a standard less than the best. It's the seeking after perfection, the striving after the ultimate, that has brought horticulture—this term in its strictest sense meaning the cultivation of plants—to such a peak. 'To strive, to seek, to find and not to yield.' One must beware of the machine.

The study of plants is a fascinating one and to succeed with their cultivation one must try to discover more about their likes and dislikes, even their limitations; try to understand their position in the plant world and, where possible, find out

I

why they have chosen a certain environment in nature. The earliest gardeners were unable to trace much of the information available today, and this was not only because the knowledge was not written down but sometimes it was misunderstood. We have to admit that this is still the case today with many plants which are most desirable and flourish in nature in conditions which would appear to be anything but ideal. Shall we ever be able to flower *Pyxidanthera barbulata* year after year in cultivation? Will *Saxifraga florulenta* ever be as plentiful in gardens as the "House-leek"? In future years will the "King of the Alps", *Eritrichium nanum*, succeed as Myosotis does in spring, and will some new understanding of the needs of *Kelseya uniflora* mean that one day it will grace a crevice in everyman's rock garden? Only more diligent observation of the plants and their responses to certain treatments will ever make these hopes come true.

Today it is easy to travel to distant lands and, once there, it is often possible, even with encouragement of a certain kind (chair-lifts, etc.), to climb and search among rare plants. As the world becomes smaller, and this unfortunately is happening all the time, conceivably there may come a time during man's evolution when the rarest plants will be in cultivation and not in nature, nowhere being inaccessible. By then the enthusiast will have to be willing and competent to accept the responsibility of keeping alive many of the plants considered treasures today.

The title of this book is *The Peat Garden and its Plants*. You now have the results of my efforts in your hands. The reason why I decided to undertake its writing was that this type of gardening is very popular today, particularly so among growers of dwarf species, and has never been, to my knowledge, the subject of a book. I am conscious that a number of small articles have appeared from time to time in horticultural literature but never anything of a sizable nature. At the outset, therefore, it might be appropriate if I were to state my objectives. These are simple. They are to bring this type of gardening to the notice of as many readers as possible, to publicize this relatively recent and very acceptable way of cultivating certain plants, to demonstrate to the gardening enthusiast with only a small area of land that he can cultivate as varied a selection of plant species as many an owner of broad acres ever did and, finally, to marshal a comprehensive list of plants suited to this type of gardening.

A plantsman, in his desire to succeed with the cultivation of a new plant, will endeavour to learn all there is to know about the plant community with which it is associated in nature. There are certain associations ranging from arctic tundra, predominantly moss and lichen, to the humid tropical forests abounding in rich lush vegetation. Since the theme of this volume is restricted to hardy wild-occurring plants in the main, it is particularly the northern heathlands, the alpine pastures, the Himalayan alpine areas, the Japanese temperate woodlands and the American temperate zones that capture our interest, with just a few dwarf species from the antipodes. On inspection, our British heaths will be found to hold a very wide range of plant types, annually new stations being found for certain species some of which were, until recently, known to occur only in one or two districts. While Calluna and Erica are dominant genera in some areas, a more mixed, yet closed, community appears to be stabilized in others. These heathlands, many of which are extremely acid and boggy, are well worth botanizing. In some instances the peat formed is many feet deep and, in some localities, is still cut for fuel. The profile of a peat bank, too, is worth looking at and it is interesting to observe the depth to which the roots of plants can penetrate into the medium. One can see the results of thousands of years of bog from the spongy, open, fibrous top layer, well filled with plant roots, down to the greasy, black, fibreless morass-like substance that exists a little way below the surface. Due to the action of frost, old peat banks are slow to become rehabilitated with plants because erosion in those exposed sites is continuous. Often it is only the level bottom that greens-up quickly. Much information can be gleaned from these sites and most make interesting studies.

1 HISTORY

Sporadically, peat gardening has been referred to over a period of almost fifty years, although in the early stages its potential was not fully appreciated. In fact the extent to which it can be developed in small gardens has yet to be realized.

One of the first references in literature where the words 'Peat walls' were used to describe a garden feature is in an article on Logan Gardens, near Stranraer in Scotland, published in 1927 in the Journal of the Royal Horticultural Society. It was written by Mr Kenneth McDouall, a keen cultivator of exotic plants, who, along with his brother Mr D. McDouall, owned the gardens. Principally the walls were used to support terraces of soil in which dwarf rhododendrons, at that time recently raised from seeds sent home by Forrest and Farrer from China, were planted. In addition some lilies, primulas and meconopsis were found sites among them. There is no mention made of any attempt at actually growing plants in the faces of the walls. The author of that article writes: 'The peat walls seem to keep the borders moist and at the same time give sufficient drainage.' Apart from these, other walls were built, this time constructed with turves. As the garden was on a slope, and in places stony, the object of building these turf walls was to retain richer soil and, on the terraces formed, to plant large-leaved rhododendrons as well as a wide variety of other exotic genera. Particularly prominent among these were species from South America, New Zealand and Australia. In time these grew, flowered and seeded, and as conditions for seed germination were ideal it was not long before their progeny were evident. This may not seem unusual but when one inspected these turf walls and determined the number of different species which had found suitable growing conditions in the faces it became apparent that here was something which could be expanded. This time, however, instead of leaving to chance the species that would colonize these walls the faces could be planted with desirable genera. One must not forget that the climate in the Mull of Galloway is very favoured and plants succeed there out of doors which would require glasshouse protection almost anywhere else in Scotland. In many ways the garden flora at Logan compares with that at Tresco Abbey in the Isles of Scilly. The winters are not cold, although salt-laden winds can bring trouble, the growing season is a long one and plants grow at a rate which is well above average. Because of the rare and choice genera which flourish out of doors there, the gardens were visited by botanists and enthusiasts. It was after an excursion to that paradise garden that senior members of staff at the Royal Botanic Garden, Edinburgh, came together and made plans. While a large rock garden has been in existence there for a number of years, the feature seen at Logan, although virtually a rock garden built of peat blocks, seemed to provide an out-of-doors environment as yet not available at Edinburgh. Also the staff of the Royal Botanic Garden were keenly enthusiastic about the same kind of plants, indeed they were committed to growing a very large collection of Chinese plants, so it was decided to construct a garden of peat blocks. On this occasion, however, the plants to be grown were not to be restricted to a few genera. The project succeeded well beyond the wildest dreams of the instigators; now it not only contains plants from the Himalayan Range but species from Europe, New Zealand, South Africa, Japan and America are also included.

Instruction by example is the ideal way to teach. As this experiment was so successful, and the Royal Botanic Garden is a public one, numbering its annual visitors in hundreds of thousands and being well known for its comprehensive collection of Asiatic plants, it was inevitable that this peat garden development should receive a great deal of attention. In more recent times many famous gardens have included some form of peat gardening in their expansion programme, and among them are the Savill Garden at Windsor Great Park, the Royal Horticultural Society's garden at Wisley, Kew Gardens, the University of Liverpool Botanic Garden at Ness and the Valley Garden at Harrogate. Although it could be said that peat gardening had its beginning in the garden at Logan, it is interesting to relate that in recent years a new peat garden styled on the Edinburgh one has been constructed at Logan, and in it a comprehensive range of plants is now accommodated.

So much for the larger gardens where staffing, while at times difficult, is at least possible, but up and down the country many smaller gardens possess and maintain small peat features. Some may be of blocks of peat built into walls, independent of any other feature, whereas others are incorporated in rock gardens, the peat blocks replacing areas of rock in the construction. In other cases sunken or at least low-lying beds filled with peat blocks have been formed while, on more than one occasion, stone troughs placed in cool shade owe their moisture-holding qualities to the pieces of peat which are used in place of rock to break up the flatness. Peat gardening has certainly not reached its peak, and every year gardeners are becoming more aware of its value.

What is it, then, that makes some gardeners want to construct a peat garden, however small? Why, too, if they want to grow small plants, are they not satisfied with a traditional type of rock garden (if there is such a thing), and why, once they have built it, does it become a thing on which they focus a great deal of interest?

In answer to the first question, that is the desire to garden with peat, there is no doubt that the peat garden provides a new set of conditions. 'Unique' may be too strong a word to use but the moist peaty mixture does provide a much sought after environment seldom met with in a rock garden whether in sun or shade. Provided

5

the peat blocks are not allowed to dry out, certain plants that sulk or grow slowly in other sites revel in the medium of the peat garden. The rock garden is ideal for certain types of plants such as those requiring maximum sunlight, sharp drainage and often a stone over which to tumble or, at least, to grow against. Most saxatile plants enjoy sending their roots down behind the rocks or, if the rocks are small, they relish creeping through the crevices. This type of association is almost non-existent in the peat garden. In the first place there is no such thing as a rock or stone over which a plant may hang nor are there restricted root runs which help many alpine plants to flower and assume compact interesting shapes. The blocks of which the peat garden itself is built are of a partially composted material and, in a way, are nature's own soil blocks. Once planted in these, certain species will rapidly send out runners and underground shoots or stolons. Theirs is not to flourish around a supporting buttress but to grow through it and, with their travelling, searching roots, actually knit the blocks together. It need not be long ere a row of sterile-looking peat blocks is transformed into a closed community of shrubby ericaceous genera. And then, instead of a hard and sometimes unkind or featureless piece of rock work, one is confronted with banks and walls of attractive and interesting vegetation. The third reason why the enthusiast for small plants gives so much attention to his peat garden is because of the extra plants he can grow in this new environment. It would certainly be wrong to say that the cultivation of the plants grown there would not be possible without a peatery, but in many instances it would be correct to say that these new conditions produce better plants. Furthermore, as the peat garden devotee is almost without exception an ardent rock gardener, his sortie into the realms of peat gardening is to provide conditions for plants that are not responding to his normal alpine culture. By providing a little extra moisture, a little extra coolness and an unimpeded root run, he sees and appreciates an improvement in his standards.

As has just been stated, the rock gardener will almost automatically gravitate to this type of gardening and this kind of keen grower is just as essential for the plants' well-being. He is, after all, the designer, owner, cultivator and menial labourer. In fact it is his skill and attention to detail, it is his concern for the welfare of his plants, his keenness, interest and knowledgeable observation that make the garden live. He must be able to recognize his plants when they first appear—apart from the shrubby kinds many will be bulbous or herbaceous—he must be diligent in his desire to remove weeds but he must also be patient and particular in his removal of them. Many shoots will be delicate and easily broken or bruised when they first become obvious and the carelessly wielded hand-fork can ruin for a season, if not for ever, some vulnerable treasure just emerging.

The peat garden may also be located in quite a small space; no need for broad vistas to enjoy the plants of this feature nor is there a need for an army of gardeners to maintain it. Rather, this is an intimate garden, the expression of one or two people. An intrinsic part of them and their close affinity to the plants will be seen in the standard of plant management they set. Theirs is not a bed that once planted can be left for the handy-man to maintain. It is an effort so personal that without their vigilance it would soon deteriorate into a feature of doubtful value.

With regard to siting, the north sides of buildings, the shady sides of shrubs and small trees (although not beneath their canopies), in some cases the not too bright areas in the garden, for example a difficult north lying slope, the base of a north wall, or even a moist part of the garden can be considered. So many plants thrive in these partially shaded areas that the problem is not what to include so much as deciding what to leave out.

2 WEEDS AND WEEDING

At first glance this may appear to be an odd title for a chapter, but there is much more to weeds than simply pulling them out. Great advances have been made in the discovery and manufacture of chemical weedkillers, but some when applied show no degree of selection between one species and another, and effectively annihilate all vegetation with which they come into contact; some also are residual, which means that certain properties remain in the soil and prevent other weeds from growing but, unfortunately, these build up to toxic levels. Most of the latter are now discarded. Another type effectively kills top vegetation but on reaching the soil immediately becomes ineffective and harmless. These may have their uses commercially, where a nurseryman, with a view to cutting the cost of maintenance, may choose to apply them instead of regular hoeing. They are valuable and beneficial when used in this context, but no one can ever say the result is tidy, clean or presentable. In fact it is most untidy, unsightly and strongly suggests slovenliness. If it were not for the competition for moisture between the weeds and the plants the weedy border would look much more natural and be preferable.

Selective weedkillers, too, have been produced, and for the seeker after a healthy weed-free greensward they are invaluable. But the chemical weedkillers, inhibitors and selectors are almost out (note I say 'almost') so far as the peat gardener is concerned. His choice of plant material is so complex that a computer would have to be on hand to help to work out a programme of control. Even this would be bound to fail because of the unknown tolerance of so many decorative plants. Also, since all would be at differing stages in their growth cycles, one would be at a loss where to begin. It may appear that only hand weeding is left, but that is not wholly accurate. If we critically define weeds, they are not restricted to locally abundant members of the native flora. They are not plants which find their way into our gardens solely by accident. In fact, in some instances, they may not be plants which are unattractive. The most precise definition is that they are plants growing in the wrong place. Some may be handsome exotic primulas, some may be the seedlings of surrounding trees, others may be the proliferation of attractive bulbous species, but, regardless of terminology, if they rapidly dominate an area and in fact show signs of suppressing less robust plants, they must be classed as weeds. Provided these are the types of weeds with which one must contend, their control, if not complete eradication, is not too difficult. There is, however, a race of weeds, the pernicious, creeping, stoloniferous kind, which quite quickly make weeding a nightmare. In the peaty medium their underground shoots travel great distances before appearing above the surface. Also, if they once get into the solid peat blocks, their control by hand is virtually impossible without the complete dismembering of the block. Quite apart from the peats, if these weeds enter the closely knit root systems of the dwarf ericaceous plants, Rhododendron being a first-class example, their removal is almost impossible. It is in such cases that one may revert to the use of hormone weedkillers carefully applied to the foliage of the weeds with a brush. In a small peatery this need not be too laborious and if the life of a twenty-year-old Rhododendron is at stake, or if the complete removal of the weed from a peat block will prevent it from invading smaller plants less able to compete with it, the result is well worth the effort. There are other plants, however, which can hardly ever be classed as weeds, Menziesia and Pieris being two which come to mind; on occasion they appear spontaneously in the walls or in the heart of a rare plant. Certainly they must be removed, but where digging them up would obviously disturb the settled walls, or where trying to save them would mean the root disturbance of a rare plant or one slow to re-establish, obviously the answer is to use a pair of secateurs and simply cut out the offending seedlings.

There is a further group of plants which can be called weeds in the peat garden. Classed here are all these invasive species, no matter how attractive they may be, which are so persistent and dominant that they constitute a nuisance to their neighbours. The degree of tolerance is determined by the vigour and detrimental effect of the invader and examples of both kinds are to be found in *Trientalis europaea* and *Uvularia sessilifolia*. In the first instance, the tiny underground stems of the Trientalis will harm nothing in their way and I know of no one who would object to dwarf rhododendrons having this plant as ground cover. The Uvularia, on the other hand, is in a completely different category. Its underground shoots, emerging from deep within the soil, spread quickly through spongy cultivated compost and grow to twelve inches or thereby in height. These have a smothering effect which would quickly eliminate much more desirable species. Therefore the aim should be when introducing a new plant to determine its habit and, if in doubt, put it in another part of the garden. If, on the other hand, one receives a plant under a very desirable name and it turns out to be a spreader, remove it speedily before it gets among the roots of the better plants and always look askance at a gift from a friend unless it has been observed growing in his garden. This is especially important if one is presented with a woody plant in the roots of which other unidentified roots are present but as yet dormant. The owner of a peat garden has to be particularly vigilant, as the conditions are so conducive to the rapid spread of stoloniferous weeds that one should look for another place for any plant that spreads unseen beneath the surface of the soil. Much unnecessary handling and transplanting will be avoided and sad heart-rending prevented if a constant watch is kept on these points.

9

3   ENVIRONMENT

A peat garden provides a new challenge to the gardener, almost a fresh dimension to his types of environment; another micro-climate which he can use to advantage. Plants are extremely variable and so are their requirements, and any new and original set-up, and one which has been proved over the years, must be welcome. It is all the more acceptable when one realizes not only the quantity of different kinds of plants that can be accommodated, but also that there is an improvement in their quality. They include a great many highly decorative species which, in some cases, are difficult to cultivate in any other way.

Wet peat certainly does add something to the atmosphere, for in the vicinity of this material it is moist and cool. Many dwarf species, certain alpine and some woodland ones, revel in this freshness and the way in which they respond pleases the gardener. Even the enthusiast feels the difference from other corners and sites in his garden and recognizes the potential. This garden feature can be looked on as another form of alpine culture but, instead of using rocks to provide features, peat blocks are used to form terraces as well as the medium for growing plants.

Far be it from me to discourage a keen plantsman from attempting to grow dwarf peat-loving plants, but it would only be honest to say that peat gardening in the sense referred to here is not for the garden owner who tills limy soils. I know some individuals who do just that and yet also manage to cultivate to a remarkably high standard a few lime-hating species, but the effort is usually very great and involves collecting lime-free water and using a lime-free compost. In the open, where the intention is to build a garden using peat blocks, the area may be free of lime for a time but it will not be long before limy conditions once more prevail. Even by building high above the surrounding soil level so that leaching from higher ground is avoided, the very nature of the structure, because of its free-standing elevated position, results in over-rapid drainage and drying out. This in turn means that resort to the district water supply is necessary and, as this is likely to be limy, the results of artificial irrigation are obvious. More than one large garden and many small ones situated in limestone areas have practised this type of horticulture but few, if any, claim success for long. Initially there may be encouraging signs, but, alas, they are usually only temporary.

The gardener who cultivates neutral or acid soil is favoured and he is free to plan and become enthusiastic over the thought of a peat garden. Fortunately this is not the type of garden that a novice might consider making at the outset. If small plants appeal, then the rock garden is usually the first choice. While I would not suggest that the peat garden is something above the level of a well-maintained and tastefully planted rock garden, it is not until the grower becomes more knowledgeable, perhaps even after admitting to a number of failures, that his thoughts turn to alternative methods of cultivation. It can therefore be assumed that the peat gardener is an enthusiast, and it is almost certain that he has learned something from his earlier efforts and is no longer a beginner. He will accordingly approach this challenge in a new light, bringing his experiences to bear when planning the new area. His knowledge of plants, too, will mean that the choice of genera, species and varieties will be a considered one.

It has already been stated that gardening with peat can be done in a number of aspects, but it is obvious that the objective is the provision of an environment with cooler summer temperatures in addition to an acid medium. So, then, the partially shaded site is probably the best suited. The more retentive of moisture the soil is, obviously the less additional watering will have to be done although, just as obviously, a waterlogged and permanently wet area should be avoided. If this is not feasible and the soil tends to remain wet it is important at the outset to provide some form of drainage. Clay tiles laid in trenches and covered with shingle or some alternative open aggregate can help. It is important that the drains have a slight run on them, the minimum recommended fall being 1 in 100, but in addition they should have a free outlet. If this is into a ditch so much the better, but remember that the outlet should be above the level of the high-water line and should face downstream so that water never actually flows into the pipe. The aim is not only to prevent flooding by keeping water out but to prevent solid matter from being washed into the pipes and, in time, choking them. If no ditch is at hand drains can still be laid, but here they should be led into a prepared sump or soak-away which is clear of the cultivated area.

One other point which should be considered if there is a choice of site—and this is a luxury available to very few—is the selection of an area on a slope and not at the lowest point, which could be tempting in a garden composed of light soil. Although many plants flower in late spring and summer, and not a few in autumn, a number bloom in early spring and many start into growth in that time. Now the low area, whilst moist, can often be a frost hollow, and this should be avoided even in a small garden, for damage by spring frost can be devastating to otherwise hardy plants.

Neither does an overhead canopy of leaves provide the right conditions, for while shelter is of great benefit to tiny plants the drips from trees spell disaster. On the other hand, shade from the side or from a distance can be good, particularly if it also provides shelter from cold, drying winds. The searing winds of March and April can do a great deal of damage to some of the more desirable semi-woodland species, and closeted nooks and crannies have to be fashioned to accommodate these special plants.

Spring droughts, too, can create havoc, so a good lime-free water supply should be available. In the small garden a few cans of water should suffice, but in a larger one some form of mobile sprinkler may have to be employed. Not only does the atmosphere in the vicinity become dry, but the peat blocks themselves tend to dry out and, in extreme cases, can qualify to be classed as fuel. Keeping the peat watered and moist is one of the operations of maintenance. Light sprinkling as opposed to hosing should be the aim.

PEATS AND THEIR SUPPLY

Before physically starting to fashion a peat garden a source of peat blocks must be found. Like so many other horticultural sundries these can vary, ranging from small peats as used for burning to much larger blocks which may be heavy to handle.

It is better to buy the blocks wet, for once they are dry the rewetting of them is a slow, tedious business. On a small scale the procuring of wet peat blocks may not be feasible, in fact if only a few peats are wanted the buying of those offered for fuel may be the only way of obtaining them. It will then be necessary to put them in a tub of water and keep them submerged by placing stones on top until they are thoroughly soaked. If a larger peat bed is contemplated, it may be possible to come to an agreement with a local supplier to provide freshly cut wet peats. The best alternative, especially if the aim is a peat garden of some proportions, is to obtain permission to cut one's own peats. The chances are that bought peats will be machine-cut and so tend to be long and narrow, like building bricks; where the size is to one's own specification then obviously larger blocks are preferable, because they make for a more stable construction. A good size is a twelve-inch cube, which should now read a 30.5 centimetre cube, or 28,372 cubic centimetres. If one is allowed to choose one's own peats, once the shrubby surface vegetation is stripped off it is the top layer, containing most of the root fibres of the plants in the heath, that makes the best blocks (see Pl. Ia). Another advantage is that one need not store the peats while the earlier work is going on, but can cut, collect and transport them as they are needed. An indisputable fact is that the more the peats are handled, particularly large wet blocks, the more they are in danger of breaking up, and as there is nothing to be gained by storing them, the minimum of handling will see most peats arrive on the site fit for use. There is no easy way of obtaining this raw material, but no special tools are needed. A good spade and digging fork and a hand barrow are probably all that is required apart from transport. The spade will skin off the surface vegetation and cut the block. The fork will lift the peat on to the hand barrow and after that it is a case of manhandling it across the heath to the road and transport. It is excellent exercise!

CONSTRUCTION

The aim of the peat garden is to create the illusion that one is gardening in a heath rather than among rocks. The areas given over to rock work in a rock garden are written off so far as planting is concerned, but this does not apply here: all the peat garden is plantable. The impression one is trying to give is of a fissured and gouged moorland where erosion and subsidence have left behind walls, banks and level terraces. The planting should be in the same vein and one should be able to visualize the plants growing right up to and through these banks, colonizing them and contributing to their stability.

When the construction of a rock garden is under consideration it is wise to seek a few lessons from nature and visit a local hillside where outcropping rocks are to be seen. By observing a few closely associated outcrops and by making rough sketches it is possible to create artificially what takes place in nature. In many ways one's efforts will be determined by the size of the pieces of stone that can be handled and by their shape. Even although a set balance and arrangement may be planned some slight modification or adjustment may have to be made because of the building material available. Certainly rocks can be moved as often as is necessary without their suffering in any way, until an acceptable pattern is reached and the rocks are bedded in.

A visit to a local moor is just as important to the peat-wall gardener. Here he can inspect nature's way of clothing the ground and observe where creeping or upright plants appear to grow best, so appreciating the value of plant association. Unlike the rock gardener, who is guided in his construction by the shape of the rocks at his disposal, the peat garden enthusiast is not restricted in this way. His raw material is malleable and he can design his peat bed exactly as he pleases. In the first instance he need only draw the outlines of his banks and slopes in soil, adjusting the heights and hollows at will, until the whole layout viewed from all directions conforms to his wishes.

During this upheaval there will have been an opportunity to get rid of perennial weeds, though this task should have been made easier by the deep and thorough cultivation that soil moving makes inevitable. One cannot overstress the need for clean cultivation at this stage. If this medium of gardening has been planned for some time and the site earmarked for it, a weed-killing programme could have been carried out previously and selective weedkillers applied according to the weed species involved. But usually there is very little time lag between the planning and constructing of a garden feature. The creative instinct does not allow the luxury of long meditation, nor is it really necessary when the aims are clear and the desire to provide a suitable home for plants is strong.

If the peat garden is to be extensive, obviously some sort of economy may have to be made in the use of peat. It may be done by simply building the supporting walls with peat blocks and incorporating in the terraces and slopes a number of blocks, buried apart from their tips. Granulated peat may then be added to the rest of the soil area in which primulas, Nomocharis, trilliums, Meconopsis, gentians and the like may be planted. On a smaller scale, where a bed is to be given over to peat-loving plants or where a sheltered north-facing pocket in the

13

rock garden is selected, a complete layer of peat blocks may be laid either prominently or sunken so that only the tops are visible at ground level. This will be governed by the rainfall in the district. Certainly one should never leave peat blocks high and dry where wind and sun will desiccate them. They should be in a

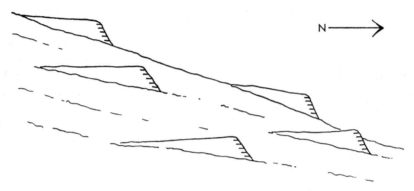

N ⟶

Figure 1

To illustrate the casual supporting of a slope using peat blocks rather than the artificial method of building solid walls or straight terraces.

position where the surrounding blocks or soil will keep them moist. In dry areas one may have to resort to supplementary watering, but this should only be necessary for a few weeks in spring—usually in April, May and part of June.

It is only after the garden has been satisfactorily shaped in soil that the peat blocks need be brought to the site. The peats can now be laid in, and if they are the fuel-type sizes they should be laid lengthwise into the soil bank to ensure maximum stability. To obtain height they may be built in overlapping layers in the same way as with bricks. The probability is that a narrow strip, lacking in volume, would soon crumble and possibly dry out owing to lack of contact with the bulk of the soil behind. In the case of larger blocks, no more than two or perhaps three layers should be required. It is important that moisture be retained in the blocks if plants are to root into them successfully, and they must be secure if underground stems are to be given a chance to spread and colonize. Make sure there is a slight 'batter' on the walls, that is to say they lie back as though leaning on the medium behind, as opposed to falling forward and thereby leaving a gap between the back of the peat blocks and the soil, which would increase in dry weather. However, the loam should never be allowed to dry out to an extent which permits this to occur. High walls are more difficult to construct and support than low ones, but some parts measuring two feet or more and facing north are invaluable for certain genera, of which Shortia and Gaultheria are two. With manageable, cuttable, shapeable peats no concessions need be made and the original design can be followed to the letter. One's artistic flair can run riot and long or short walls may be included at will. Naturally they will not be straight, like those built of bricks, but can be quite informal, being deepest where the front protrudes prominently, gradually becoming more shallow as they recede and finally disappearing where the last block merges with the soil. From the highest point of the garden, however, there should be a naturally flowing slope which by terraces and banks eventually ends at the base of the garden.

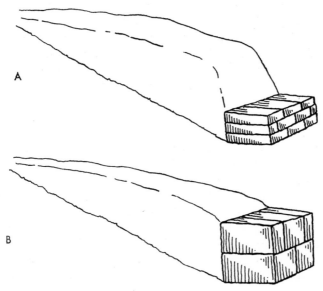

Figure 2

Methods of laying peat blocks against the soil bank.
A. Using narrow peats similar to those cut for fuel.
B. Using larger blocks cut to specification.

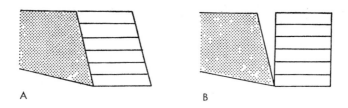

Figure 3

Relation of peat blocks to soil bank.
A. Peats leaning backwards, the rear of the blocks in contact with the soil. This is the recommended method.
B. Showing what can happen between the soil bank and peat blocks if the peats are allowed to dry out.

Just as in the rock garden, it is important that no air spaces be left when building is finished, and all the odd corners where these could occur should be filled with a mixture of soil and peat as work progresses. At the end, when the last peat is in position, a liberal dressing of granulated peat should be spread over the soil and be worked into the surface layer. The amount will vary according to the vegetable matter already in the soil, but, as a guide, where the soil is almost purely mineral 50 per cent of the total bulk should be peat.

This is the time to apply a balanced fertilizer, and any of the proprietary brands will do, but only half the recommended quantity of dressing should be used. Many small plants are unhappy in soils containing concentrated fertilizer, the autumn-flowering gentians being an example, but they respond to a mere dusting. Therefore, just like the convoy system where the speed is determined by that of the slowest ship, so the dressing of fertilizer should be reduced to a level which will not harm the most fastidious of plants.

STEPPING-STONES

The rock garden has an advantage over the peat garden in that the stones also provide places where one may stand without harming the plants. No matter what size the rock garden may be, it should never be necessary to stand on the soil in order either to weed it or carry out any other maintenance such as pruning, top dressing or transplanting. A suitably placed rock should always be there. In the peat garden a few stepping-stones should be incorporated. Without being either obvious or large they should be of sufficient size to enable the site to be tended without treading on soil. If the garden is successfully planted there will be no bare soil on which to stand, as the whole area from the peat banks to the level terraces will be full of dwarf species. Standing on them would cause havoc and leaving bare soil would just encourage weeds, so the placing of stepping-stones in the site solves all these problems. Small pieces of broken sandstone paving are probably less conspicuous than other types, and certainly formal slabs would be incongruous. As the soil or medium in the peat garden is so spongy and open, treading when weeding would make the ground so compacted that not only would it be necessary to loosen the soil with a digging fork, but the texture, too, would be ruined.

4 PLANTING ARRANGEMENT

The peat garden is now ready for planting—but what kind of plants will thrive best in these conditions? Many shrubs, as well as herbaceous and bulbous plants, will respond to this environment and the first types to plant should be the woody ones. These are the permanently visible inhabitants and so the correct placing of them is most important. Some are evergreen, others are deciduous, some are tall and upright while others may be short, sprawling or creeping. They all have more than one function to serve. In the first case they must be decorative. There is no point in including a plant that is not handsome in its own right, for, unless one is deeply interested in making a collection of a certain type of plant, the less attractive members are better left to botanic gardens. Next, they can be used as a foil or background for some other species or cultivar planted in front or, in some instances, be used as supports for certain delicate climbing plants such as *Codonopsis vinciflora*. Apart from these considerations the shrubs can also provide shelter. The main point is that they can serve as a 'bield' (shelter) from the cold drying winds of early spring which can do so much harm to precocious plants and some not completely hardy evergreens. Dwarf plants positioned around and among them can be protected from wind or, in some cases, frost, if tucked underneath.

Obviously the larger shrubs are for the background. Because of their isolation, i.e. the individual planting, all the specimens will tend to have foliage to ground level. To dispel any suggestion of grading, an occasional well-proportioned plant can form a feature near the front of the peat garden. The pattern of their siting should be anything but formal and this is best achieved by placing them in an organized yet seemingly haphazard way so that the spaces left between them vary all over the garden. At the outset there may be a tendency to overplant with shrubby specimens because of their stature at the time of initial planting. This can be a mistake and within a few years lead to trouble unless one is resolved to remove the surplus plants before they harm the shape of their neighbours by excluding light by being too close.

It is better to err on the sparse side with planting, and rely on shorter-lived fillers until the permanent plants take over. Obviously the size of the peat garden will determine the ultimate size and number of shrubby species to be planted, but, just like any other type of planting scheme, most of the smaller plants must finish up at the front. Tall-growing Meconopsis and lilies may be planted some distance back where their roots and bulbs are in shade while their flowers are in the open. Certainly the more interesting non-invasive type of herbaceous perennials of medium height will flourish in many of the terraces, while evergreens of a sub-shrubby nature may find a more acceptable position in the shade at the base of one of the peat walls. This is a garden where it is possible to display one's plants in a natural setting and so form plant associations which in nature would not occur.

There is no area of the peat garden which need be purely functional—a dreadful word to use when describing a garden and its plants. Everything is of interest and should never cease to cause comment. While it can be said that certain garden features may be furnished with plants of wild origin, all garden areas owe some-thing to the hybridist or observant plant selector. This corner, as much as any other, provides the environment for a host of nature's wildings.

After considering all the factors, a simple order of planting could then read as follows: taller shrubs in greater density towards the back or distant periphery of the garden; in the middle distance less tall shrubby species should be included, and in the more open spaces herbaceous perennials such as primulas, orchids, thalictrums, trilliums, etc.; while the foreground (apart from an occasional well furnished woody plant which would help break the force of the wind) would be colonized by dwarf, creeping, sprawling, suckering genera of which typical examples are gentians, miniature primulas, ourisias, omphalogrammas and cassiopes.

5 SOIL FIRMING AND PLANTING

After the soil has had time to settle, and just before planting begins, the whole area should be lightly trodden. As it will tend to be moist it would be a mistake to trample the ground too firmly, but most of the plants would not flourish in too loose a soil, and light overall treading on a dry day when the soil is not too wet should provide the right compression. With the correct quantity of peat added the soil should break down naturally. A general levelling with the digging fork will also ensure a professional finish. A continuous process of levelling as work progresses will retain an even structure in the soil and avoid the mounds and hollows which can result from planting specimens with large root balls. Notice that the tool suggested is a digging fork and not a steel or spring rake. The last two tools produce a tilth fit only for seeds and, if used, will provide ideal germinating conditions encouraging weed seeds to grow and flourish.

Shrubs, of course, must be firmly planted, and this can be done at any time during the dormant season, but at the end of a day, and if the soil has been in the right condition for planting, a good watering will help to settle the soil around the roots. This will carry small soil particles into spaces that may have been left unfilled. The plants will become established more quickly if this is done. More robust herbaceous species may be introduced during their dormant season and, like the shrubs, most will appreciate firm planting. Bulbous kinds, too, have their more varied times for lifting and replanting and at the appropriate season they should be allotted their spaces. 'Allotted spaces' means just that, and a sensible selection where one neighbour does not influence another to its detriment is important. One must allow for seasonal spread; more than one dwarf shrub of many years has been ruined because a smothering type of herbaceous perennial has been allowed to become too rampant.

Most of the smaller, and in many ways the more interesting, plants, however, are better left until spring. It is at that time, when soil lifting by frost action has ended, that they move more successfully. Many of the dwarfs, particularly those which have a shallow root system, are extremely vulnerable, so that the period between their transplanting, which is when they are still dormant, and the time root action begins should be as short as possible. Careful watch must be kept over soil moisture lest it falls below an acceptable level. These tiny shrublets, of which the plant family Ericaceae contains not a few, need careful planting. The finished soil level should be just above the topmost roots and no more, and the degree of firmness such that the plants remain steady. Just the correct amount of pressure has to be applied so that the tender roots are not damaged by crushing.

A method of planting often seen is one of digging a hole, placing the plant in the centre and filling in the soil, firming all the time in a manner that suggests potting up. Now this kind of approach may be the right one when the plant has a large amount of fibrous roots to which much soil has adhered, as in dwarf rhododendrons. The correct planting depth may not be so critical that the exact soil line of the plant should correspond with the level of the surrounding loam, but, because of this approximation, an uneven soil level could result. Most other plants, however, of which primulas, gentians, ourisias, saxifragas, clintonias are examples, have different root systems and a satisfactory way of dealing with them is to draw out a suitably sized hole with the trowel and place the young plant against the undisturbed back of the hole. The existing soil level and the appropriate planting depth of the plant can be accurately compared and adjusted in a trice. The soil is then returned to the hole and gently but firmly pressed against the plant. There is then no necessity to knead the soil behind the plant as this has never been loosened. The pressure of the hands rather than the trowel is more easily assessed and also more efficient when you are dealing with these plant treasures. A quick sweep of the soil by hand is all that is required to produce a level surface. Finally, before turning your back on those recently transplanted, give a generous watering to settle the soil around them.

CULTIVATION

On the whole, since the plants in the peat garden are perennial, there will rarely be a need to cultivate the soil deeply. Generally, weeding and tidying up are continuous as all manner of debris, broken twigs, leaves, etc., keep getting lodged in among dwarf plants. Nothing is shown off to advantage in a messy situation, and it is only by keeping the garden clean and tidy that one will obtain full satisfaction. Apart from the hand fork, which will be as closely attached to the peat gardener as a glove, a pair of secateurs is very much part of his equipment. Broken and dead branches and those which spread excessively will have to be removed when they are seen. This is all part of plant maintenance, a phrase suggesting machinery in a factory, but which relates to something just as important in the garden.

If seeds are not required it is unnecessary to tax a plant by leaving it to produce excessive quantities. Therefore flowers may be removed as soon as they are spent so that the plant can divert its energies into producing growth rather than seeds. In this way young plants are encouraged to add girth to their stature and as a result look more healthy and robust. They will not have that dry stunted look so apparent after heavy seed bearing. Not only shrubby plants suffer because of this kind of neglect; many herbaceous perennials do likewise. If the production of seeds is discouraged there will be no great germination of unwanted seedlings, which can be as much of an embarrassment and nuisance as weeds; many primulas can be placed in this category. What is more, these self-sown seedlings may appear in the centre of a precious perennial, they also compete with the introduced plants for food and water and, apart from influencing them in this way, they can spoil their shape and often cause them to lose their lower leaves due to overcrowding.

Another operation that ought not to be neglected is top dressing. This can be done at any time of year, but is best in spring just as growth begins. The object of top dressing plants will be quite plain to those who have handled these miniatures. The fine fibrous roots of many of the sub-shrubby types will be laid bare through the removal of weeds and by soil erosion. The effect of erosion will be very obvious if watering by artificial means has had to be carried out, as water applied in this way is never as gentle as rain showers. The herbaceous and dwarf plants will also benefit from top dressing, for some of these, certain woody-based primulas being examples, tend to grow out of the soil and become stunted in a year or two. Regular top dressing, to replace the soil removed or lost in this way, will help to prevent this. Dwarf shrubby plants which increase in girth by runners, stolons or trailing side shoots, examples being Andromeda and Cassiope, will respond in an amazing way if the bare parts of their shoots are buried in soil. This will help them to increase more quickly, for uncovered shoots will never root because of movement due to wind. Also, that part of the stem which remains uncovered becomes hard and woody and is less likely to produce roots. Dry stunted shoots result. One might ask why this sort of treatment is so necessary in what we term almost ideal conditions when obviously, in nature, the plants must fend for themselves and certainly can never expect a regular supply of well-prepared compost. In the wild state, however, the plants are not subject to a form of monoculture: they have associates. Because of the unnatural situations in our gardens the companions they would have would be grass and luscious weeds rampant enough to swamp them and rob them of their food and moisture, and competing with them for light. These are not the types of plants with which these treasures have to contend in their native homes. As we have observed before, the balance struck in these closed communities, of which the heathland is a very good example, allows all species to thrive to a degree. If we refer back to our Andromeda and the side shoots by which it expands (although these are not top dressed in the way we think of when we use that phrase) these shoots are covered by a canopy of other vegetation. It could be "Tormentil" or even sphagnum in nature, or a dwarf mossy saxifrage in the garden. The main point to note is that these side shoots are soft, etiolated and moist, are in a perfect rooting medium and so become established quickly.

All this cultivation and maintenance do encourage our plants to increase in size. Initially when they are tiny and without character their growing brings satisfaction and enjoyment; later, as they start to outgrow their allotted space, they themselves constitute a problem. The obvious answer with shrubby species is to transplant, and if a group of plants were used to provide density the removal of half or even two-thirds of them might be necessary to bring the planting once more into scale. Ultimately, after a number of years, only one specimen may be needed to fill that space. The important thing to remember is that woody plants should be thinned out in good time, that is to say before the plants touch each other and start spoiling and ruining their shapes. This process may have to be performed every two or three years, depending on the plants' growth rate. The obvious alternative is to ignore the plants altogether and permit them to grow in and through each other. This may be all right in some instances, especially where plants had been given a generous allocation of space at the outset to allow them to make good-sized root systems. When closely planted, however, the competition tells on the less well established individuals. Just when the canopy appears to be completely closed and the ground perfectly covered, one or two of the weaker specimens either die and leave holes in the vegetation or they cease to put on much growth. This causes the mound of foliage to look patchy. The transplanting of the dwarf herbaceous plants and bulbs should be just as diligently carried out, for it is a mistake to wait until a plant is impoverished by overcrowding before moving it. By then one is actually moving a second-rate plant, and at least one season if not two may elapse after replanting before it recovers. It is far better to keep plants healthy and robust and, when a species is known to multiply rapidly to its own detriment, it should be transplanted at regular intervals while it is still in prime condition. The notebook, not the memory, should be used to remind the gardener of moves to be made. Only then will the plants be dealt with in the correct or most appropriate season when they will suffer least.

LEAF-FALL OR AUTUMN MAINTENANCE
Interest in plants is retained throughout the year because of the changes that take place all the time, especially if they are noted for their autumnal colouring. The plants may be late summer or autumn flowering, as certain of the Asiatic gentians are, but more often it is fruit or foliage patterns that one finds attractive in those seasons. Many of the latter are to be found among the peat-loving plants. Attractive though these changing leaf colours are, in the majority of cases the leaves belong to deciduous shrubby plants or trees. In their own way they decorate the garden but, while the rustle of walking through a carpet of leaves is a delightful sound, fallen leaves lying for days on dwarf prostrate plants can do a great deal of damage by denying light to certain plants, whose shoots become drawn, especially those of small shrubby plants. They can also smother the display of a patch of autumn flowering plants, and if they are wet and soggy they may cause some of the more vulnerable to rot off.

The removal of the leaves as they fall is a constant job, for many of the established plants are of the type that object to this kind of covering. This may seem a long-drawn-out affair, but frankly, once the leaves start to fall, the chore is very soon over. An early frost is liable to produce heavy leaf-fall immediately after-wards, but in effect this will result in a less prolonged leaf-gathering season.

Evergreens, of course, are part of any garden. They are very important and reliable plants, but it is unfortunate that their leaf-shedding time does not coincide with that of deciduous plants. Evergreens cast most of their unwanted foliage in spring and early summer, so, if too many of the bigger leaved ones are included in the planting plan, leaf cleaning will have to take place at that time. They are usually not so serious or damaging to the plants for two reasons. One, they are firmer and do not pack down on dwarfer plants in the same way as the others do and, two, at that time of year conditions are usually much drier, so there is less fear of their becoming soggy and rotting, thus spreading this malaise to healthy tissue.

Often the evergreen species are already part of a planting scheme, as the decision to use this area for a peat garden may have been taken because of the number of shrubs established there. Their presence can be beneficial to the other species planted in their shelter or shade, but as this part of the garden is to contain lots of treasures and, in consequence, will have to be handled frequently, there are certain kinds of shrubby plants which are quite unacceptable. Better a formal fence or a young, more suitable, species than established Berberis or holly. Spiny plants should be banished from the vicinity where hand cultivation is to be enjoyed.

6 PLANT TYPES

If one were to make a list of all the plants used in peat gardening it could be very lengthy. Because of the relatively low rainfall, east coast gardens would include a great many species which would normally grow happily in less exacting surroundings in the west. Actually, due to the wetter conditions in the west, a more open and sunny situation might be necessary to make the plants bloom. In the south, where summers are warmer and the dew (which in effect is as beneficial as a damping down), is evaporated or absorbed by breakfast time, plants may be placed in the peat garden simply as a means of providing cool conditions in the shade. In the north the grass may remain damp until mid morning. So this 'peat garden' idea is not completely restricted to a few fussy plants in the same way as one might think of specimens for an alpine house; the choice of plants is extensive, and the combination of so many beautiful and less common plants growing happily together creates a wide appeal.

Obviously it would be wrong to say that all plants could flourish in peaty conditions. If, however, one analysed the plants mostly used and segregated them into their plant families a majority would undoubtedly fit into three: Ericaceae, Primulaceae and Liliaceae. Included in these groups could be many hundreds of different plants. Within Ericaceae alone, which in the main is woody or subshrubby, would be found an ample number of evergreen genera and species. Many have highly decorative flowers while others may be grown almost solely for the attractive fruits they produce in autumn. In range, the plants vary greatly in height, spread, type of foliage and size of flower, and only the dimensions of the garden would determine their choice. Ericaceae is the "Heath" family and embraces some of the most attractive dwarf shrubs grown today. There can hardly be a garden which does not grow some member of this very large family. The bulk of the species will not thrive in limy soils, due no doubt to the absence of mycorrhiza which plays a very big part in their intolerance of certain growing conditions. Ericaceae is a family about which most growers of alpine plants are enthusiastic. It is a very large one, embracing more than 70 genera and 1,500-odd species. Add to these the many subspecies, wild varieties, the endless proliferation of cultivar names, and the figure becomes astronomical. The thought of 1,500 species alone seems formidable, but when one considers that the genus Rhododendron itself claims more than 600, Erica, after which the family is named, 500 or thereby and Vaccinium over 130 then, obviously, many genera must contain very few. This is so. In fact a lot of genera are monotypic, that is to say they possess only a single species which, it could be said, should make identification easy. This, however, need not be the case. Take, for example, Calluna, the "Scottish Heather" or "Ling". *Calluna vulgaris* is the only species, but when one considers the number of wild collected forms and the varieties selected from the stock beds of nurserymen, one is immediately aware of the tremendous potential of a single species. The identification of the various forms of heather growing in our gardens today is a full-time study. Owing to its wide interest and appeal a society was formed, called the Heather Society, which aims at bringing together all those who are interested in heaths.

Ericaceae, then, includes many individuals. Some are tall, as in *Arbutus menziesii*, which is recorded in nature as growing to a height of 100 feet or more. This is a handsome evergreen tree from California with attractive peeling bark reminiscent of Eucalyptus. *Rhododendron arboreum* is another member which is tree-like in its measurements and one which, when mature, flowers extravagantly. At the other end of the scale are plants of minute stature suitably illustrated by *Loiseleuria procumbens* and *Harrimanella (Cassiope) hypnoides*. These are subarctic and mat-forming in nature.

Plants belonging to Ericaceae are found growing at sea level in the subarctic regions of the world, but as one moves into the more temperate zones the environment at that altitude changes, and the conditions favoured by these plants are then found only on the higher hills and mountains.

Not all ericaceous species are hardy out of doors in Great Britain, so it is also true to say that, so far as our climate is concerned, some members of the Ericaceae are greenhouse plants, but this is of little consequence. Truly this family has a worldwide distribution; both hemispheres contain plants which belong to it and, while those from south of the equator may have an added appeal to the armchair traveller, many of those native to our own northern regions are extremely decorative.

Obviously it would be both enjoyable and enlightening to deal with this plant family in great detail but, while it is a very important one when classed with our main subject, to treat it thoroughly would require a volume to itself. Even eliminating the larger species, for the very good reason that they are out of scale in the peat garden, we can still list more than sufficient to satisfy the plant collector.

On the whole, ericaceous plants prefer a moist, lime-free soil. Some species are shade tolerant while others, particularly some of the miniature alpine denizens, must have maximum sunlight. Without doubt the amount of light influences the formation of flower buds, but in our gardens it is more often a matter of available moisture that determines a plant's position in the garden rather than shade. This is a factor extremely important to these shallow, fibrous-rooted plants.

Not only do the species in Ericaceae differ in height, leaf shape and flowers, but there is a wide range of fruits which in itself is fascinating. There are berries as well as dried capsules. In many cases the former are strikingly attractive and are the main reason for growing the plants; in this category one could include *Gaultheria miqueliana* and *Pernettya mucronata*. Or berries may be delightfully edible, as in *Vaccinium oxycoccus*, the "Cranberry".

Propagation is by seed, layers, grafts or cuttings, depending on the species involved, but generally speaking increasing stock is not difficult.

Plants bloom for a certain length of time, but for the greater part of the year no flowers are present. It is important, therefore, if one is to get the utmost from a plant to be able to appreciate it outwith its normal flowering season. Ericaceous plants are of that calibre. They have personalities, or it might be better to say they possess attractive characteristics which they display throughout the whole year.

Another very large plant family is Liliaceae, and among its members are a great many which fit ideally into the surroundings of the peat garden. Unlike the ericaceous plants, which on the whole are shrubby, the genera included in Liliaceae are mostly herbaceous. Many of them are bulbous by nature, although there are exceptions, examples of which are Philesia, which is shrubby, and Liriope and Helonias, which are evergreen and so permanently visible. Their distribution ranges over all the temperate and subtropical zones of the world, and many of them are extremely attractive. This is a very large family in which there are about two hundred genera listed and ten times that number of species. Hybridists have exploited this family in no small measure and, while many wild species are in every way as beautiful as any man-made hybrid (and not a few enthusiasts would shout 'more so'), there is now available a very large number of garden hybrids, selections and strains. Indeed, taking the genus Lilium as an example, it is now very difficult to procure the true wild species. It is fortunate that in every continent there are groups of plant enthusiasts whose concern and interest lie in the world's native flora. These observant and dedicated amateurs and professionals conserve in their gardens and for our pleasure many of the world's natural treasures. That this interest is appreciated and infectious can be seen by the number of gardeners with the smallest of plots who cherish these rarities of nature. One frequently hears of difficult, rare and unusual species being cultivated, but old hybrids, having served their purpose and become either exhausted or superseded, are discarded like old boots; the wild, garden-worthy species, the very root of our interest, are never spurned in this way, and if lost to cultivation are searched for and their absence regretted until they are reintroduced.

Members of Liliaceae are monocotyledons, identifiable by having the parts of their flowers in threes or multiples thereof. There are a few annuals which on the whole are best ignored, and there are some tenacious perennials, not least among them being certain species of Allium and Ornithogalum, which are better shunned like the plague. Certainly if they ever became established—or would it be more correct to say naturalized?—in a garden, they would torment the cultivator for the rest of his life. Despite these few less attractive members, no gardener could spurn the truly beautiful Nomocharis and Lilium, many of which have their homes in the Himalayan valleys. No shallow rooting plants these, as the majority send their feeding roots deep into the medium in search of food and moisture and, while Ericaceae is noted for its fibrous root system, Liliaceae has bulbs, thick fleshy roots or thongs. If their basal roots are perennial the transplanting can be a hazardous operation, and the re-establishment of disturbed plants will not only take longer but can, by the poor standard of growth and flower, be glaringly obvious during the first year. While the moving of plants should always be done efficiently and thoroughly no plants reward the careful gardener more for his prowess than those which are liliaceous.

This very large and worthy plant family contains enormous variety. Apart from annuals there are short-lived or long-enduring perennials, species which are bulbous, rhizomatous, tuberous, stoloniferous, herbaceous by nature, evergreen in some instances, shrubby and climbing in others and, as though to emulate the orchid, in more than one instance epiphytic.

Much can be made of all these attributes in the peat garden, and they bring to the area an air of grace and beauty that is surpassed by no other group of plants.

The third large family, which in no small way contributes to the interest, colour and beauty of the peat garden, is one to which one of our most attractive British native plants belongs, the primrose, and so to Primulaceae. Unlike the other two, however, which contain a large number of genera as well as species, the Primula family, while being variable and embracing approximately five hundred species, is restricted to fewer than thirty genera. Most of these are found in the northern hemisphere and include highly attractive members. Because of their known requirements, not a few of them seem destined for the peat garden. Primula is the genus which immediately comes to mind, and so accommodating is it that larger gardens finding room for a few more vigorous kinds (and in a much less pretentious setting) can be swamped by the numbers of self-sown seedlings. Apart from Primula, which contributes a great deal to the charm of the garden, Trientalis ("Chickweed Wintergreen"), Cyclamen ("Sow-bread") and Dodecatheon ("Shooting Stars") can be considered for the various aspects. The fascinating task of fitting in the numerous genera, species and varieties is a most satisfying exercise and one which demands a great deal of knowledge, interest and enthusiasm, but with rewards so great that even the beginner, provided he selects sensibly, can reap a worth-while return.

To draw a line here and say that a peat garden devoted solely to plants listed in Ericaceae, Primulaceae and Liliaceae could be a complete garden would not be inaccurate, for surely three hundred genera containing four thousand species and involving many thousands of wild varieties and cultivars would supply all our needs. But man is not like that; he is not built to be satisfied with a ready-made restricted set of plants. He is by nature an experimentalist, and if told of these limitations would immediately set out to prove the statement wrong. It is just as

27

well that no such restraint is imposed, for this book takes account of close on another thirty plant families and, when considered in full, the choice becomes almost limitless. The only bounds are those set by the garden fence and the collection determined by the space available. It would perforce be duplication to list all the plant families dealt with, but Ranunculaceae, Saxifragaceae, Berberidaceae, Orchidaceae, Gentianaceae, Bignoniaceae, Polygalaceae, Polygonaceae, Diapensiaceae and Celastraceae may serve to reveal the vast range of plant types which follow. The majority are selected for their ability to flourish in the environment just described, and it is guaranteed that all who profess to be plant conscious will find in this setting a wealth of plant material exceeded in no other type of garden feature.

7 THE HEATH FAMILY (*Ericaceae*)

ANDROMEDA is a monotypic genus, circumpolar in its distribution. It is part of the vegetation of boggy heath and grows in close association with other plants. In fact its method of growing—spreading underground or at least through the lower, moister plant layers in a soil rich in humus—should be copied in cultivation; its long shoots are best covered. Because of its wide natural spread in the colder areas of the northern hemisphere the species varies in the shape of its leaves and flower colour, and not surprisingly names have been applied to these forms.

A. polifolia is an evergreen shrublet up to eight inches in height which carries its flowers in tight clusters. These flowers would be completely round if it were not for the small opening and the minute reflexed tips of the petals. On the whole they are pink. (Pl. 2.)

A. polifolia var. *angustifolia* has narrow leaves and is one of the taller growing varieties.

A. polifolia var. *compacta* is close growing with shoots which are less inclined to wander. Its dense habit makes it a worth-while plant, and a white form sent from Japan in recent years is even more attractive. The foliage in this instance is distinctly glaucous. It is usually listed as 'Alba' or (White Form) to distinguish it from the variety.

A. polifolia var. *macrophylla* is the form with the broadest leaves. These are distinctive, and one plant in particular displayed its travelling habit when surrounded by a mat of *Saxifraga cespitosa*. The shoots popped up among the rosettes of the saxifrage and were vigorous and healthy so long as they enjoyed this association.

A. polifolia var. *nana* is the name under which the most popular form is listed. It is a compact plant and although stated to be dwarf, five inches or thereby, need be no smaller in stature than the preceding variety.

ARCTERICA is another genus which can claim only one species. In the past it has been known as *Pieris nana* and *Andromeda nana*. This is the "Northern Heath", and is native to north-east Asia and Japan. It allows its buds to open in April, although they are visible from the previous autumn. Creamy-white is the colour of the tiny lily-of-the-valley type flowers.

A. nana is the specific epithet, but 'minuta' might have been a more appropriate name as this tiny evergreen barely ever exceeds one and a half inches. Its tenacity in spreading through a spongy peat block is revealing, especially after it has been observed growing in an ordinary lime-free soil mixture. The very small leaves are hard and tough when mature. (Pl. IIa.)

ARCTOSTAPHYLOS is the "Bear Grape" and is a fairly large genus including many species, but only a few are considered here to be of suitable size for the peat garden. All are confined to the North American continent, apart from one which has a wider distribution. The species are evergreen and the little bell-like flowers are carried in terminal clusters and are always fairly prominently displayed.

A. columbiana, although eventually a plant of taller dimensions, is well worth considering because of its attractive foliage. The leaves are pale green at first, darkening with age, but it is their downiness when young and the dense spreading canopy they form which give the plant appeal.

A. × media is one of those fortuitous natural hybrids which occurred aeons ago, and is now considered to owe its existence to the ability of *A. columbiana* to cross with *A. uva-ursi*. The dwarf stature of the last mentioned is fairly dominant as the height of this hybrid rarely exceeds eighteen inches. It forms an effective ground cover and, while young and vigorous, sends its long sweeping branches across the soil or through neighbouring plants. The lighter green of the young foliage contrasts well with the dull darker shade of the mature leaves. The small flowers which appear in April and May are pale pink.

A. nevadensis is mat-forming and is by many considered to be a mountain form of *A. uva-ursi*. It is prostrate in habit and varies very little from our native species.

A. uva-ursi is the wanderer, being found in all northern continents, and is an attractive British native plant. It can be seen clothing some banks by the side of the main Perth–Inverness road. While other plants are present between the clumps this prostrate shrub effectively suppresses any plant over which it spreads. Again, while the small pink flowers are attractive, it is the variation in the foliage shades which is so pronounced and interesting. Eventually one plant may be measured in square yards, but there is no difficulty in confining it to a limited area in the garden.

BRUCKENTHALIA is very much like a dwarf heath when not in flower. Its tiny foliage, less than a quarter of an inch long, is so dense and the shoots so numerous that no soil is visible through its vegetation. There is only one species, and at one time it was known as *Erica spiculifolia.*

B. spiculifolia has a height of around six inches, on top of which the terminal spikes of deep pink flowers with prominent stamens are displayed. Both spikes and flowers are numerous and effective. June and July constitute its flowering period. Top dressing encourages the shoots to root so that old specimens may be dealt with as herbaceous plants and be divided and replanted.

BRYANTHUS, like many members of the Heath family, has appeared under a number of names and has also had included in it many species which are now dispersed through other genera. Today's classification recognizes but one species.

B. gmelinii, syn. *B. musciformis,* has long been known to alpine enthusiasts, although it can only be looked upon as a plant to complete a collection of dwarf ericaceous genera rather than as an attractive addition to the garden. It is an interesting plant, none the less, being completely prostrate, having the tiniest leaves and numerous interlacing shoots, but, alas, I have yet to see it flower. Apparently it does have flowers and these are recorded as being pink and borne in a few-flowered raceme. Its derivation, mosshead or mossflower, clearly indicates its size.

CALLUNA is a wonderful genus which is much admired on the mountain as well as in the garden. It grows wild on our highest moors, and while it is affection-ately known as "Scottish Heather" it is met with in eastern North America and in western Asia. It varies greatly in the shade of its flowers, the colour of its foliage and its habit of growth. Although only one extremely variable species, it has a potential that hardly seems possible from seeing it on the hillside.

C. vulgaris is the "Ling" or "Heather". It is said to be lucky to have a sprig of white heather and now, because of selection and observation, white forms may be had throughout the whole of the Calluna's flowering season. Many new cultivars are available to gardeners, and although some of those more recently offered have occurred in nurseries the older variations were introduced from the wild. Even the very popular 'H. E. Beale' with its long spikes of double, silver-pink flowers was a chance find in the New Forest. The coloured foliage forms have been much admired in recent years, and just like most hardy coloured foliage plants the shades are more intense in winter. Those which display yellow or golden leaves must have maximum exposure to light and a peaty soil. To list the many cultivar names would simply duplicate readily available catalogues, but for those wishing a late flowering white, *C. vulgaris* var. *hirsuta* forma *albiflora* is one that can be recommended.

Remember that pruning the old flowering shoots in February and top dressing with peat will not only help to keep the plants vigorous and within bounds but, in addition, guarantee long sprays of flowers.

31

CASSIOPE may be divided into two sets: the Sino-Himalayan group to which *C. fastigiata, C. selaginoides* and *C. wardii* belong, and the circumpolar group of which *C. lycopodioides, C. mertensiana* and *C. tetragona* are well-known examples. None is really difficult to grow, although some succeed better than others, and not all flower well in some gardens nor do they flower well every year. They vary in stature from prostrate, spreading individuals to completely upright specimens as erect as the traditional Irish Yew. The more light these plants have the more flowers they are likely to produce, but, depending on the garden and the nature of its soil, a little shelter or even shade may be necessary to prevent the plants from drying out in summer. All have evergreen imbricate leaves and flower during April and May.

C. fastigiata is native to the north-west Himalaya and has leaf margins which are ciliate. There is a furrow down the back of each leaf and this, too, is fringed with hairs. The stems of this species are very stiff and the individual white flowers are borne in the axils of the upper leaves. *C. fastigiata* has been collected from various localities, and while some survive not all forms appear to be hardy in our gardens. (Pl. 6.)

C. lycopodioides favours cold arctic conditions and, in keeping with plants found in exposed areas, has a prostrate habit and forms mat-like clumps. It has tiny foliage, but no groove divides the back of its leaves, and when it is flowering well, no vestige is visible of the foliage or stems.

A top dressing with fine soil to which peat and sand have been added is beneficial to the plants if applied after flowering. North-west America and north-east Asia are given as its natural distribution and it was first discovered in Kamchatka more than two hundred years ago.

C. mertensiana is found in western North America. It has a rather loose prostrate habit and could never be termed good ground cover. It blooms sparsely and, like the previous species, has leaves which have a rounded keel.

C. mertensiana var. *gracilis* is a much more closely knit, dwarf, wild variety. This is assuredly a good garden plant, being a little more upright than the more widely grown prostrate species and, when in bloom, there is no confusing the larger white bells of this plant with any of the other forms. It is restricted to Montana and Oregon.

C. saximontana is a dwarf shrub up to eight inches high. It is occasionally listed as a subspecies of *C. tetragona*, a plant which inhabits the arctic regions of North America and Europe and is a difficult one to cultivate. *C. saximontana*, on the other hand, is confined to the Rocky Mountains. It flowers well in an open situation and tends to have a yellowish-green hue when compared with the darker green of other species. (Pl. 7.)

C. selaginoides is a most attractive species native to the Himalaya where it was collected a number of times. The stems are upright and on these are borne the comparatively large white globular flowers which are held well clear of the foliage on long ciliate pedicels. This is particularly true of the form introduced by Ludlow and Sherriff in 1950 under their number L & S 13284. Plants of this particular gathering received an Award of Merit when exhibited at the Royal Horticultural Society's Hall in 1954. (Pl. IIb.)

C. *wardii* is the much sought-after giant among the cassiopes. It has a loose habit of growth, the lower parts of the stems often being covered by brown dead leaves while the upper parts carry in four rows the green to grey, tightly overlapping foliage. The colour varies according to the form, some being much more hairy than others. Plants almost twelve inches high have been recorded.

In recent years, through the raising of plants from seed saved in gardens, a number of hybrids have appeared. Some were distinct enough to warrant cultivar names, but, as with many popular plants, the proliferation of names tends to confuse rather than sustain interest in a genus. Among the popular varieties is C. 'Edinburgh' (Pl. 5), considered to be an offspring of C. *fastigiata* and C. *saximontana*, although it was from seeds obtained as C. *mertensiana* that it was produced. It is a truly magnificent plant when in flower, especially when young, as older specimens tend to be straggly and less attractive. As it is easily raised from cuttings the aim should be to keep up a supply of young plants. C. 'Muirhead' claims C. *wardii* and C. *lycopodioides* as parents, and this is one of the finest dwarf hybrids available. The influence of C. *wardii* is very easily seen in the hairy leaves and the firm stout young shoots. Other names are used to identify specifically selected clones, but as the species cross freely one must beware of applying too many names.

CHAMAEDAPHNE is another monotypic genus closely allied to other genera, differentiated by minute botanical characters. It has long been known to gardeners as Cassandra and was first recorded well over two hundred years ago. Its wild stations are to be found in northern Europe, Asia and North America.

C. *calyculata* is a shrubby species of up to three feet or more with long arching branches on which the oval leaves are alternately arranged, but in such a way as to appear to be folded back. They almost form a line along the top of the shoot. Hanging singly from the axils of these leaves are the white globular flowers. C. *calyculata* 'Nana' is smaller than the type in every way and is no doubt a selected seedling chosen because of its dwarf stature.

CHIOGENES is one of those ground-hugging genera, in this case containing only one species, which never fail to fascinate the plant enthusiast. It is closely allied to Gaultheria and Vaccinium, in fact it is sometimes included in the former with the specific name of *serpyllifolia*, resembling the Thyme. So prostrate are the thin stems and tiny leaves that they have virtually no vertical measurement.

C. *hispidula* from North America bears the common names "Creeping Snowberry" and "Ivory Plum". The inference in both cases is that the fruits are white, which indeed they are, being in fact berries.

33

CLADOTHAMNUS is yet another North American monotypic genus in the family. It is deciduous, and although considered a tall shrub is slow growing and flowers regularly from a small size.

C. *pyroliflorus* tends to have shoots which are upright and branching and these carry smooth glossy leaves an inch or more in length. The pink flowers have yellow tips and are attended by green sepals, the whole making a most interesting combination.

DABOECIA is extremely popular with those involved with heath gardens, where it can usually be found; both the time and method of flowering make it desirable in other places. One species is found wild in Ireland, in fact there it was christened "St Dabeoc's Heath". It carries its flowers in long narrow racemes well above the narrow evergreen foliage and each flower hangs gracefully from a long slender stalk.

D. *azorica* is the most widely sought, being less than nine inches tall and producing the deepest of blood-red flowers. The shoots tend to be twiggy and in the colder parts of the country it cannot be considered hardy. Invariably the plants grown under this name are hybrids and no doubt their hardiness is due to the influence of the other parent, D. *cantabrica*.

D. *cantabrica* is certainly the more reliable member and in an ideal situation may exceed two and a half feet in height. Pruning in early spring keeps the plants within bounds. Because of the wide colour range it is not possible to give a specific colour to the species, but numerous names indicating the shades have long been applied. The names 'Alba', 'Atropurpurea' and 'Rosea' are self-explanatory, while 'Bicolor' identifies a form with three kinds of flowers, some purple, some white, while others are both purple and white, a most unusual mixture.

ENKIANTHUS is one of the most distinctive of genera, and once a person has been introduced to its finer points it forever holds great appeal for him. The species are deciduous, some being quite massive finally, and the method of forming their narrow branches in tiers, and the singular way in which the short lateral shoots appear, producing their leaves in whorls, give the whole pattern a graceful appearance. Add to this the thousands of delicately held drooping clusters of tiny bells and one has a plant of great merit. To ensure further that its merits may not be forgotten too soon the foliage in autumn takes on the most brilliant colours before being shed.

E. *cernuus*, a name well known to shrub growers, is a Japanese native species and will eventually reach five feet in height. It is the variety E. *cernuus* var. *rubens* which should be procured, however, as its stature is much less and it never fails to produce flowers of a rich red colour.

E. *perulatus* is a most meritorious species, having the purest of white flowers. These are produced just as the leaves start to unfold and are virtually precocious, so that the plant, covered at this stage with white orbs, looks most attractive. In autumn the narrow leaves change from green to deepest purple, passing through all the intervening colours and shades before falling.

EPIGAEA has long been of interest to the connoisseur. It contains three species which certainly bridge the gap between East and West. All are prostrate sub-shrubs, bearing thick leathery leaves on thin wiry stems bristling with hairs. They are not the easiest of plants to grow in Britain; they can sometimes be decidedly difficult, for the reasons that they are hard to obtain in the first place and equally hard to retain. They enjoy the protection of other plants and prefer to grow in shade under a leafy canopy in a sheltered site. And to the grower who successfully flowers these well must go the cultivator's award. Their flowering season extends from March to May.

E. asiatica flowers in May and carries its deep pink blooms in few-flowered racemes. The flowers are tubular with the petal tips reflexed. It is native to Japan and is reputed to be less difficult to grow than its close relative *E. repens*.

E. gaultherioides is a fairly long name for a dwarf and difficult plant. It is not an easy plant to produce, for seed is the only sure way of increasing stock and, as one must be a successful grower to flower this plant and so obtain seeds, it is not very often that spare plants are available. The first flowers usually open in March. It comes from the Black Sea district of southern Russia and Turkey where it grows in association with *Rhododendron ponticum* and *Vaccinium arcto-staphylos*, forming ground cover under them. (Pl. IIe.)

This plant has been known for many years as *Orphanidesia gaultherioides*, but within recent years has been included in Epigaea. It is quite distinct from the two other species, however, as here the flowers are large and open and do not possess the long tube of either *E. repens* or *E. asiatica*.

E. repens, sometimes known as the "Mayflower" of America and sometimes "Trailing Arbutus", is a collector's plant, native to eastern North America from Canada to Georgia. This creeping evergreen shrub, although hardy, obviously misses the snow covering it gets at home. The British winters encourage the young buds to break too soon, with the result that they are frequently cut by frosts. Like *E. asiatica*, this species carries its flowers in clusters, but here they are much narrower and a paler shade of pink.

E. 'Aurora' is a hybrid between the Japanese and American species and in its characters is more or less midway between the two parents. The flowers are certainly darker than those of *E. repens*.

ERICA is the genus after which the family is named and includes a very large number of species, many of which are attractive, although, alas, only a few are suitable for planting out of doors in this country. Included in the hardy ones are many decorative wild-occurring cultivars and some which have appeared spontaneously in nurseries and gardens. Like the Calluna, or "Ling", a catalogue of names could be compiled, but so many are already well known and appreciated that very little further reference need be made to them.

As sometimes happens, certain members of a genus may choose to grow in acid peats while others may be found growing in certain limestone formations, but as the latter kinds show no reluctance to flourish when planted in peaty soils no omissions need be made here on that score. In the same way as the callunas are cut back so should the ericas be dealt with. Pruning after flowering, in some instances severe pruning, may be necessary to restrict their spread. The flowering period of the European mat-forming *E. carnea* is so long that no excuse can be found for leaving it out.

E. carnea—by some now referred to as *E. herbacea* and by others more affectionately as the "Mountain Heath"—has central and southern Europe given as its natural distribution. Although this latitude could suggest tenderness it is completely hardy in our climate. It was introduced into our gardens more than two hundred years ago, since when it has been popular and useful. In time it will form large mats of vegetation which are transformed into the most colourful carpets during the flowering season, which is winter and spring. There is wide variation in the flower colour, foliage shades and time of flowering, and if a careful selection is made one can have colour for months on end. The method of flowering is in the form of a compact terminal raceme and its ability to smother weeds effectively is not the least attractive of its attributes.

E. ciliaris, the "Dorset Heath", cannot claim to be as garden-worthy as *E. carnea*, but its habit of growth and leaf arrangement are more or less peculiar to itself. The leaves, occurring in whorls of three, are prominently glandular and hairy. It is a plant which appreciates moisture, especially when young, and at no time should ever be allowed to become dry. It is a long-lived plant, in fact plants forty years old can bloom just as well as those of a few years. Its flowering period is autumn and the large bright flowers, drooping on a terminal raceme, produce a bright display.

Although there are a number of forms, the two most desirable, apart from the species, are *E. ciliaris* 'Aurea', and *E. ciliaris* 'Stoborough'. The former is shy to bloom but bright with golden foliage, a useful characteristic, particularly when it does not lose a great deal of its colour in partial shade. *E. ciliaris* 'Stoborough' is one of the more robust and produces white flowers.

E. cinerea is the species, along with *Calluna vulgaris*, which can claim credit for brightening the hills in summer. It is, of course, the "Bell Heather", and unlike our last species will flourish in the poorest of soils. *E. cinerea* has a long flowering period extending well into autumn and, in addition, because of its widespread distribution which covers the whole of the British Isles and extends into much of western Europe, it shows great variation in flower colour. This has resulted in numerous cultivar names being given and, next to the "Ling", it is probably the most widely planted. Golden foliage plays a part in the appeal of

1 *Adonis vernalis* (page 104).

2 *Andromeda polifolia* (page 29).

3   *Arisarum proboscideum* (page 105).

4   *Cardamine asarifolia* (page 109).

5   *Cassiope* 'Edinburgh' (page 33).

6   *Cassiope fastigiata* (page 32).

7 *Cassiope saximontana* (page 32).

8 *Cornus canadensis* (page 112).

9 *Cypripedium calceolus* (page 115).

10 *Disporum smithii* (page 84).

11 *Erythronium revolutum* (page 85).

12 *Fothergilla monticola* (page 120).

13   *Gaultheria rupestris* (page 39).

14   *Glaucidium palmatum* (page 125).

15   *Ledum groenlandicum* (page 43).

16   *Leiophyllum buxifolium* (page 43).

17   *Leucothoë davisiae* (page 44).

18　*Lilium martagon* (page 90).

19 *Lilium canadense* (page 88).

20 *Loiseleuria procumbens* (page 44).

21 *Meconopsis discigera* (page 130).

22   *Meconopsis latifolia* (page 131).

23   *Meconopsis regia* (rosette) (page 131).

24 *Meconopsis* × *sheldonii* (page 130).

25 *Meconopsis simplicifolia* (page 132).

26   *Menziesia ciliicalyx* (page 45).

27   *Mertensia virginica* (page 132).

the species and, like many others, in addition to varying in flower colour, plants of differing heights can be obtained.

E. × *darleyensis* needs little introduction to gardeners. It is fairly vigorous and spreading, but these two not altogether unwanted factors may be kept in check by pruning immediately after flowering. Certainly it would be criminal to avoid mentioning this bispecific hybrid which is noted for its long flowering period. It first appeared at Darley Dale in Derbyshire many years ago, but since that time other seedlings have emerged more noteworthy than the original. The parentage is *E. carnea* × *E. mediterranea*. One with deep pink blooms has young growth which is yellow and brightens up the whole planting; another, a German introduction, bears white flowers which remain fresh for months.

E. × *darleyensis* 'George Rendall' and 'Silberschmelze' are the two cultivars specified. One useful cultivar with completely golden foliage in summer and autumn is 'Jack Brummage'.

E. mediterranea, now, I understand, a name sunk in synonymy and more correctly known as *E. erigena*, has a fastigiate growth form to which is added, in most cases, glaucous foliage. Some forms are of dwarf habit, and these are the most hardy, while others up to three feet or more in height, though ideal in the background, are not so colourful or hardy as to warrant unlimited planting. Some bays in counties Galway and Mayo and areas of the western seaboard of the European mainland are where it may be found wild. Its flower colour ranges from white to pink and deep rose.

E. tetralix, the "Cross-leaved Heath" is a plant familiar to all who tramp the hills. In some areas it may be dominant, but usually it is sparsely dispersed amongst the rest of the heath-forming vegetation. It is one of our most distinctive plants, even in isolation, as no other vies with it in its stiff habit and proudly held terminal clusters of relatively large pink bells. One could almost repeat here what has been said of the other species, for *Erica tetralix*, too, has its share of garden forms. Though this summer-flowering species has few outstanding named cultivars, one of the finest with dark red flowers is *E. tetralix* 'Con Underwood', while in 'Mollis' the blooms are white and the foliage is silver. The "Bog Heather" is another common name given to this plant, this readily indicating its need for moisture. Seedlings will appear in peat brought in from the moors.

E. vagans, the "Cornish Heath", grows wild on that peninsula. Its precise station is on the Lizard, the most southerly point of Great Britain. It is also distributed throughout western Europe. This is one of the last ericas to bloom and, when it does, its six- to eight-inch-long spear-shaped racemes are solid with flowers.

The flower colour ranges from white to deepest pink, and in choosing names for some forms reference has been made to the areas where the particular cultivar was found. 'Pyrenees Pink' is an obvious one while 'Lyonnesse' and 'St Keverne' signify the area surrounding Tintagel where the legendary King Arthur held court.

GAULTHERIA has included in it another group of species in which a great deal of interest is shown. The distribution is world-wide, for they are to be found in Asia, North and South America, New Zealand, Tasmania and Australia. Not all are hardy, of course, but there are enough which vary sufficiently in stature to make them look different. Gaultherias are attractive evergreen woody plants, some bearing colourful fruits, others being adorned by interesting foliage while others again owe the appeal to their flowers or habit of growth. Many of them are small in stature, and in this section are the New Zealand species, but as they have shallow roots one must beware lest their roots become dry. It is easy to put off too long the day when one should irrigate, so it is better to err on the moist side and apply a top dressing of peat as an added precaution.

G. *adenothrix* from Japan can be a trifle tender in an open situation, but in among other shrubs, even beneath the canopy of some overhanging evergreen, it will receive the shelter necessary for healthy growth. The growth habit is a sprawling one and along these spreading stems are produced the unpolished, almost round, leaves up to one inch in diameter. It is the flowers, however, that capture most attention. As in many of the Ericaceae, the flowers are globular and in this instance they are pure white, but this pallidity is offset by the calyx, which is deep red in colour. This combination alone would guarantee a place for the species, but in addition bright red fruits develop later in some gardens.

G. *antipoda* is said to reach up to five feet or more, but usually it is seen as a pros-trate specimen, a dwarf billowy mass of small, brownish-red foliage. This New Zealander could be classed as a foliage plant, although it does produce small white flowers. Large red or white fruits are recorded by other writers and, when these occur, obviously they provide the plant with increased appeal.

G. *cuneata* is one of the 'musts' in any collection of Chinese ericaceous plants. It is a reliable evergreen which, as its specific name suggests, has cuneate (wedge-shaped) bases to the leaves, although the points of the foliage tend to have the same outline. The foliage is hard and distinctly shows a net venation. Its flowers appear in spring borne in small terminal racemes. By mid August the most attractive fruits are evident in the form of white orbs and for some weeks are a conspicuous feature. It has to be admitted, however, that in some gardens at least the birds are very fond of these berries and, once tasted, they are very quickly stripped. The texture of the spongy peat blocks is ideal for its spread, and plants in this setting, at least forty years old, can be measured in square feet. Ernest Wilson collected seeds of this Gaultheria in west Szechuan where he said he found it in woodland and nearly always on rocks.

G. *depressa* is of New Zealand origin and in many ways is a more attractive plant than G. *antipoda*. Its growth is lighter and the leaves are tinier, all the characters being about half size. The bronzy appearance of the foliage, so prominent in colder weather, gives the species an all-year-round appeal. The minute leaves are extremely bristly. It is found in boggy areas in the wild.

G. *forrestii* is named in honour of George Forrest. When the blue fruits develop one realizes how fascinated the collector must have been the first time he saw the plant. It is a little vigorous for some gardens as the arching branches can be eighteen inches to two feet in length and along them the narrow leathery foliage is alternately disposed. While grey-green on top the undersides of the leaves are paler. Once flowering stage is reached, axillary spikes of white flowers brighten the planting. It is by suckering shoots that this species spreads.

G. *miqueliana* is the species most confused with G. *cuneata*. It is native to Japan and occurs round the margins of subalpine, coniferous forests, forming dense patches of vegetation and increasing its territory in the same fashion as G. *cuneata*, i.e. by underground travelling shoots. It is also attractive and more robust. The leaves and berries are larger, but unfortunately the berries are just as palatable to the birds. These fruits are pink-tinged, but the species is easily identified by the rounded ends to its broad oval leaves.

G. *nummularioides* could have been made specially for a peat wall. The stems are completely prostrate on level ground or are pendent when arching out into space. The bristly hairy stems make a handsome covering to the brown peat blocks. The wiry shoots carry their leaves in opposite pairs, and what makes them so attractive is the arrangement whereby the larger almost circular leaves at the base of the shoot gradually reduce in size until, at the shoot tips, they are quite tiny. Pinkish flowers and purplish-black fruit follow in season.

G. *procumbens* must be familiar to all. It is the well-known "Winter-green" or "Partridge Berry" which annually sends its colonizing red runners through many inches of peaty soil. It forms a dense carpet one or two inches high, although in the plant's youth longer, more erect, shoots may appear. The leaves tend to gather, umbrella-fashion, at the tips of the side growths and it is amongst these that the pink, usually solitary, flowers appear in May. Later the fleshy fruits, bright red in colour, make this species highly decorative. A form bearing large fruits is sometimes offered under the cultivar name of 'Macrocarpa', but among plants raised from seed some variation is bound to occur.

G. *pyrolifolia* could be termed a poor form of G. *miqueliana* when viewed for foliage effect. It has the same reticulate type of leaf, but is inclined to produce much shorter shoots. The flowers too are smaller, but what is most apparent is the colour of the berries, these being almost black. Its distribution is given as eastern Himalaya.

G. *rupestris* is yet another New Zealand native species. This plant certainly has a stiffer appearance and is much more compact and shrubby than any so far discussed. It also carries its creamy white flowers in erect racemes well clear of the foliage. Some leaves are almost round, up to half an inch in diameter, while in other forms the leaves may taper at both ends. The circular leaves seem more suited to this plant, which is compact and round in habit. The fruits are dry. (Pl. 13.)

G. *shallon* is simply included as a warning. It is one of the few plants which will grow successfully beneath the canopy of a beech tree. Be it dry or moist—shaded or in full sun—this plant will thrive. Because of these attributes it would be worth considering were it not for its invasive underground shoots. Pink flowers in a prominent raceme followed by purplish black fruits sound attractive and would be welcome if only this species were less rampageous.

G. *sinensis*, a most attractive dwarf shrub, is decorated in autumn with large bright blue berries. These are preceded by creamy white flowers in April and May. Like all gaultherias this species is evergreen. Although the shoots tend to spread rather than grow upright the habit of the plant is quite compact, for it rarely exceeds six inches in height. Seeds of G. *sinensis* were collected by Frank Kingdon Ward in Burma almost thirty years ago.

G. *thymifolia* is one of those tiny plants which, although of a woody nature, are very difficult to class as shrubs. It is really a tight miniature, spreading through the peat blocks at a perceptible pace and eventually forming a closely knit cover of tiny shoots. The small white to pink flowers tend to be hidden among the shoots, but it is not a plant that flowers regularly in cultivation. If these flowers were fertilized the resulting fruits would be blue in colour. This is one of those plants, native to the border of Tibet and Upper Burma, which are fascinating to so many people.

G. *trichophylla* is often confused with the previous species, but of the two there is no doubt which is the better. Unfortunately this species is in some ways tender but is well worth trying, for the reward to the successful cultivator is the large greeny-blue fruits which it bears in autumn. Its flowers, too, are larger and more colourful than those of G. *thymifolia* and, whereas the shoots continue to spread underground in both species, G. *trichophylla* tends to become bushy with short shoots radiating from a vertical axis. The flowers are flushed pink and are quite large in relation to the size of the plant.

G. *veitchiana* produces long arching branches which are quite bristly, while alternately arranged on them are thick, leathery, broadly lanceolate leaves. They can measure as much as four inches in length. The venation is quite prominent even on the top surface of the leaves. The shoots can be up to three feet long and then, produced in the axils of the upper leaves, come racemes of small, white flushed pink flowers. Light blue fruits follow.

The French missionary Père David first found G. *veitchiana* a century ago, but it was not identified as being new until after it had been rediscovered in Szechuan by Wilson, who was at that time collecting for the Exeter firm of Veitch.

G. *yunnanensis*, superficially, is in many ways similar in appearance to G. *veitchiana*. If anything it is stronger growing and one need cultivate only one of these species for its form since both share the same graceful arching habit. This species is less colourful in autumn, and the fruits, which are black, are the best means of determining which plant is which. G. *yunnanensis* was at one time known as *Vaccinium yunnanense*, and in literature may still be found under that name. The attractive manner in which it displays its shoots makes it a plant ideally suited for placing in a prominent position.

GAYLUSSACIA is a genus of dwarf to medium-sized shrubs native to the American continent. There are quite a number of species, but very few are in cultivation, those that are being grown for their decorative flowers or fruits, and are the "Huckleberries". Gaylussacia is very similar to Vaccinium, a genus with which it can be confused, but here the ovary is multi-celled, which immediately distinguishes it. In nature it prefers a sandy-peaty mixture with varying amounts of moisture.

G. *brachycera* is the species most likely to appeal to the peat gardener. It is an ever-green, not unlike our own native *Vaccinum vitis-idaea*, and bears the common name of "Box Huckleberry". At one time it was described as *Vaccinium buxi-folium*. This species spreads slowly by underground suckers, and in nature some extensive areas dominated by it are considered to be colonized by single plants. This would imply that individual clumps are probably thousands of years old. In our gardens, however, we can never hope to see plants quite so old, but small plants with tightly packed, robust shoots will be admired by more than one enthusiast. Continuous top dressing aids this species in its spread.

HARRIMANELLA is a small genus containing but two species. In fact it is only within recent years that they have been placed under this name; prior to that they were classified as Cassiope. In many ways they resemble that genus, but one obvious difference is in the leaf shape. Cassiope, generally, has thick imbricate leaves, while in Harrimanella they are much thinner and are held away from the stem. One could say they enjoy the same conditions, but one species, extremely difficult to grow, would have to be compared with the less easy cassiopes.

H. *hypnoides* is an arctic plant, little more than an inch high, found in northern Europe. Norway and Finland show it in their native plant records. It does not grow alone, however, and no doubt this is a clue towards its successful cultiva-tion, for in nature mosses, *Phyllodoce caerulea* and other dwarf plants are listed among its associates. Cool moist conditions are essential, but if these are supplied along with shade this tiny plant will soon die. It can be encouraged to grow or at least exist for a few years in the peat garden, but I doubt if many growers will see it produce flowers. These are small, white flushed with pink and Cassiope-like, being carried singly on thin reddish pedicels capped by red calyxes.

H. *stelleriana* is easier, having small leaves which are quite prominent and are yellowish-green. It forms a mat of vegetation and above this the small, pure white, solitary globular flowers are set off against red sepals. Light top dressing as for Cassiope suits this plant and, although recorded as an alpine on dry rocky slopes, the dry atmosphere in the rock garden does not provide the ideal conditions for it in cultivation. Neither does it want a dark area, so it should be given as much north light as possible. Its distribution in nature extends from western North America through the Kuriles to some of the Japanese northern islands where it is said to be rare. (Pl. IVc.)

KALMIA is commonly referred to as "American Laurel", but all who have seen kalmias well flowered must wonder why the ubiquitous, dull laurel was ever upgraded sufficiently to be compared with Kalmia! The genus is confined to eastern North America. The conditions most suited to this plant are those favoured by rhododendrons, although it must be said that higher summer temperatures are probably necessary to make the plant flower well. The flowers are so geometrically formed that one could not be blamed for thinking they were artificial. There are only a few species, seven being the stated number, although varieties increase the interest and choice.

K. *angustifolia* is the narrow-leaved Kalmia or "Sheep Laurel" which blooms in early summer, the flowers being produced in clusters. It is native to the north-eastern part of the continent, which almost assures its hardiness, and in our garden grows up to three feet. The flower colour is said to vary between white and purple, and cultivar names like 'Rosea', 'Rubra' and 'Candida' are applied to the particular clones. I have yet to see the white one. K. *angustifolia* 'Pumila' is of dwarfer stature with bright rose-coloured flowers. (Pl. Vb.)

K. *latifolia* is the species most frequently encountered, and when conditions are ideal it is one of the most decorative evergreen shrubs. Too often, however, it is seen in an unhappy state and consequently appears dull. It transplants easily and this should encourage the gardener to move it if the first site selected is not ideal.

K. *polifolia* is a loose-growing evergreen shrublet of eighteen inches or thereby. The dainty pink blooms are carried in few-flowered clusters, and although completely hardy, being native to Newfoundland, its sprawling habit does not produce a neat plant. However, it does allow smaller species to be grown close to its stems without influencing them to any great extent.

KALMIOPSIS is of relatively recent introduction, having been first discovered in Oregon in 1930. Its name means 'Little Kalmia' or 'Kalmia-like' and this is a very apt description. Kalmiopsis is monotypic and is a dwarf evergreen shrublet which for many years was considered to be a difficult plant. Today this is less so, although it would still be wrong to say that it could be cultivated successfully in a variety of sites. The peat garden environment has proved amenable to it— so, too, has the cold frame—but because of the shallow surface roots dryness in summer must be avoided.

K. *leachiana* bears the name of Mrs Leach who first discovered it. Of upright growth, it carries flowers on elongated racemes, and from the original seeds plants twelve to fifteen inches high were raised. At first it was thought to be of very limited distribution, but in recent years further collections were made in other parts of Oregon in the Umpqua River Valley. These have produced better garden plants, not only because of their more compact habit and larger flowers but because they are much easier to grow. M. Marcel le Piniec was the first to find a variant and, although there appears to be confusion about naming, there are two at least: one named 'Marcel le Piniec' (Pl. Vc) and the other (Umpqua Valley Form). The obvious difference between the two is in the colour of the flower. Other cultivar names have since been added. These dwarf forms grow well in the rich peaty soil, where the foliage is virtually screened behind masses of pink flowers.

LEDUM is a small group of evergreens, one of which is well worth including in any collection of dwarf shrubs. The flowers are interesting in the way the stamens protrude beyond the petals, giving a spiky appearance.

L. groenlandicum, known colloquially as "Labrador Tea", is a worth-while species and young plants can cover themselves with terminal clusters of creamy-white blooms. They flower well at an early age and can be replaced before they out-grow their allotted space. (Pl. 15.)

LEIOPHYLLUM is another monotypic genus occurring in some eastern states of North America. It has been known since 1736 and is one of the plants found in the Pine Barrens of New Jersey, where it is commonly called "Sand Myrtle". It appreciates good light, but the soil must always be moist, particularly while the plants are small.

L. buxifolium blooms in June. The flowers are pink while in the bud stage but white when fully open. Moreover, although the individual flowers are small, they are produced in clusters and in great numbers. This dwarf shrub has the tiniest leaves and on old mature plants the leaves and shoots are so numerous that the foliage presents a solid mound of vegetation. (Pl. 16.)

There is a slight variation within the species and this has been recorded in the varietal names *hugeri*, in which the fruit coat is rough, and *prostratum*, in which the leaves are said to be opposite.

LEUCOTHOË includes both evergreen and deciduous species, and these are much grown by all who appreciate graceful arching branches and decorative foliage. The tendency is for the plants to spread through loose soil, colonizing and dominating the area as they expand. They are extremely effective at sup-pressing annual weeds, although no one need take this into account when assessing a genus. All those mentioned bear white flowers and because of their spreading growth are better placed round the edges or at corners of the peat garden.

L. axillaris is an evergreen with long curving shoots, some measuring four feet or more, but as the stems are arching two to three feet is the plant's true height.

43

The shoots are well adorned with lanceolate leaves, so that it is impossible to see bare soil through the thicket of shoots.

L. davisiae, collected in 1853 in the mountains of the Sierra Nevada and named in honour of a Miss N. J. Davis, is one of this family's most beautiful plants. Once established, it spreads by suckers and effectively forms clumps in semi-shaded parts of the garden. The leaves are dark green and glabrous and are arranged alternately up the stem. Terminating these shoots, which extend to two and a half feet or more, are stiff racemes of creamy white bells. These appear during June and July. (Pl. 17.)

L. fontanesiana is similar to *L. axillaris* in many ways and is often confused with it, but in this instance the leaves are acuminate, the points being long drawn out. It is more robust, too, in the young stage, but mature plants which have removed much of the feeding from the soil assume a less vigorous habit and produce foliage which, although evergreen, is multi-coloured. It is still met with under the synonym *L. catesbaei*.

L. grayana differs from the other species by being truly shrubby, almost suckerless and with branching stems. It is also deciduous, but this gives the plant an opportunity to display its reddish shoots in winter. The leaves are ovate and in the axils of the upper ones the spikes of flowers appear. This plant is Japanese and varies slightly, some forms producing leaves which are more glaucous than others.

L. keiskei is one of the less robust species. Records state that it rarely exceeds twelve inches in height, and it has shorter branches, which are distinctly wavy. Leaves are attached at every point where there is a change of direction, giving these shoots an almost unique appearance.

LOISELEURIA is a subarctic genus and another which contains only one species. In nature it shares an open site with other dwarf heath-forming plants and mosses, often in extremely exposed places. It is truly tiny, barely rising above soil level, and it requires good light if it is to flower.

L. procumbens was long known as *Azalea procumbens*, and even today is often affectionately referred to as "Creeping Azalea". It has a wide distribution in the northern hemisphere where its diminutive pink flowers are much admired in late May. In the peat garden it can be encouraged to grow on the top of a solid block of peat. (Pl. 20.)

LYONIA is closely akin to Pieris and the species are sometimes included in it. A number are deciduous, which means that with dying foliage there is a chance of enjoying autumn colour, especially in gardens where this is generally good.

L. ovalifolia, although a large shrub, can be slow to grow, and even when small—that is, twelve inches high or thereby—it can look colourful in autumn. The foliage, too, is decorative, being oval in shape and slightly glaucous on the underside.

MENZIESIA obtained this Scottish name from Archibald Menzies, a plant collector who, searching for plants in North America at the end of the eighteenth century, discovered *M. ferruginea*. This species is sometimes called "False Azalea", but it is not among the more desirable plants. This shrubby genus is a deciduous one and found not only in North America but extending as far as Japan, where the more garden-worthy species are.

M. ciliicalyx is the finest member. It is deciduous and may reach up to three feet in height. Just at bud-burst, when the leaves are barely half expanded, the tubular, bristly, light wine-coloured flowers are at their best. They are pendulous and translucent when seen with the sun behind them; the young leaves, too, are handsome and wear a glaucous bloom. Seedlings of this Japanese species occur in the peat blocks, but it is important to remove them early, with a view either to transplanting or discarding them lest their removal at a later date disturbs the walls of peat. Certainly they may be cut over at ground level if unwanted. (Pl. 26.)

M. ferruginea, the first member to be discovered, can reach five feet in height, but it is usually much less. Its very light pink blooms are produced sparingly.

M. purpurea is another handsome plant and one that must be mentioned because of its floriferous habit. It is a more robust form of *M. ciliicalyx*, being sometimes four to five feet in height, but, like so many other species, decorative from an early age. The long tubular flowers are purplish, as the specific name suggests, and are produced in clusters.

PERNETTYA are plants which berry and retain their fruits throughout the winter. These are assets to any garden feature and are just two reasons why certain pernettyas are popular. They are not really plants for shade, although they will tolerate a certain amount, but it is in the more open sites that their fruiting qualities can be fully appreciated. Some are invasive and the ideal spongy texture of peat blocks encourages rapid expansion, therefore it is important that when making a selection this fact is borne in mind.

P. mucronata is native to South America, growing wild in the Magellan Region and Cape Horn. No doubt it was discovered by A. J. Pernetty when he sailed with Bougainville to the Falkland Islands almost two hundred years ago. It is a very variable species in both shape of leaf and colour of fruit. Many hybrids raised in gardens have been distributed and these are the ones most decorative, fruit-wise. Names like 'Alba', 'Coccinea' and 'Rosea' describe the colour of the berries, while 'Bell's Seedling', an extremely reliable fruiting form, although somewhat rampant, indicates a grower's selection. The species is of course evergreen and completely hardy, and as some forms tend to be self-sterile it is better to plant three or four together to ensure cross pollination.

P. nana is a plant that must be included in any volume dealing with dwarf ericaceous shrubs. It is one of the true miniatures, being scarcely three inches in height with wiry stems bristling with tiny, deep green, blue-tinged foliage. It is not in any way invasive; in fact it needs careful looking after and is really a species for the connoisseur. This New Zealander is recorded as having reddish fruits, but these are none too plentiful in this country.

P. *prostrata* in some ways resembles P. *mucronata* except that it does not roam, nor does it stand stiffly erect. It may reach nine inches in height, with thicker leaves than P. *mucronata*, but the fruits are as large as or even larger than the best of that species. At first a lilac shade, they soon turn to deep purple. The form known as P. *prostrata* var. *pentlandii* is said to be more upright in growth. This species from the South American Andes is not too hardy.

P. *pumila*, sometimes listed as P. *empetrifolia*, produces a dense carpet of bright green vegetation. It makes good ground cover in shade or sun, adding a change in texture to that type of situation, but in general it is rather dull. This South American species will spread quite rapidly by underground stems and by the rooting of the prostrate shoots, so that one must beware lest it swamp more precious plants.

P. *tasmanica* will always be in demand. It is one of those fascinating plants that the enthusiast recognizes in every garden where it is grown. It is still not seen very often, despite its neat habit. The stature of this Tasmanian Pernettya is dwarf, barely reaching four inches in height in shade while mat-forming in more open situations. The tough little stems bear tiny evergreen leaves in a scattered pattern and in some of the leaf axils miniature white flowers appear in May. Later the fleshy fruits develop and these can be quite large, relatively, some-times almost half an inch in diameter. In the main they are red in colour, but varietal names applied indicate that some have white while others have lemon-yellow berries.

PHYLLODOCE is sometimes referred to in old books as Menziesia or Bryan-thus. Some species have been in cultivation for almost two centuries while others are of much more recent discovery. This group of tiny shrubs is invalu-able not only because they are easy to grow in a peaty soil but because they are evergreen and are of hemispherical shape. In most instances old, well-estab-lished plants build themselves up until the twigs are so dense that the plants could pass for heaths. For healthy growth the soil should be able to retain moisture and must never be allowed to dry out. Like most ericaceous plants they do not like their stems buried too deeply; only a light covering of soil over the roots is sufficient.

Phyllodoces are confined to the northern hemisphere and found mainly in the arctic and subarctic zones. In some countries their provenances may extend southwards for many miles, especially where the plants have altitude as an alternative to latitude. Unfortunately this is one of those genera where there is much confusion. No doubt this is due to their tendency to hybridize in the field, which has resulted in crosses being introduced as species.

P. *aleutica* is a lemon-yellow flowered species native to north-east Asia and western North America. Like all phyllodoces it is evergreen and long lived, specimens thirty years old being not uncommon, and in this particular species the corollas are devoid of glands. Eventually the plants will reach nine inches or more in height, but they are exceedingly handsome and floriferous during their growing-up years. The round, almost globular, flowers, constricted at their mouths, are borne in terminal clusters. The species most likely to be confused with it is *P. glanduliflora*, but the difference is quite obvious and is explained in the specific epithet, for the flowers are glandular.

P. breweri is the exception so far as form is concerned, for instead of assuming a neat rounded shape it spreads laterally and can form a circle a yard in diameter. The long sprawling shoots are well furnished with needle-like leaves, but these become much smaller nearer the terminal racemes. These flowering spikes are up to six inches in length. Unlike those of the previous species the flowers are open and the rosy-red petals are reflexed; this in turn exposes the stamens and so gives the flowers their spiky appearance. This species is native to California, where it grows in moist places at an altitude of 6,000 to 12,000 feet. (Pl. 34.)

P. caerulea is of special appeal, for it is one of Britain's rarest alpines. For many years it was recorded in only one station, but within very recent times it has been observed in other localities, albeit within the same county in central Scotland. This plant is truly arctic and is quite abundant in the colder climates. It can be slightly straggly in growth, especially when young, but eventually it becomes as dense as the tightest of heaths. It is both hardy and reliable and will furnish an open situation as heathers do and with the same ease. The pink urceolate flowers can completely hide the foliage. (*See* Pl. Ib.)

P. caerulea var. *japonica* is a variety that has long been cultivated, and although it is listed as 'hardier' that scarcely seems the correct term to use. 'More amenable to cultivation' might be better, since the hardiness of *P. caerulea* cannot be in question. This variety, however, is of dwarfer habit and tends to form a mat rather than a hummock. The interlacing branches never seem to knit so tightly, in fact the whole plant is of looser habit and the foliage is darker. Apart from these differences the flowers are darker in colour, being a shade of bluish-mauve, while the form of the corolla is almost inverted pear-shaped with a small opening. (Pl. XIc.)

P. empetriformis expands its few-flowered clusters of rosy-purple blooms in May. Again the shape of the corolla is the main guide to the species and the word used to describe it is campanulate. There is no constriction whatsoever and each is formed like a bell with the tips of the petals turned back. The plant often grown in gardens as the species is one of the selections from the naturally occurring hybrids, and in fact is *P.* × *intermedia* 'Fred Stoker' (Pl. 36). Empetrum-like foliage is not a good description for two reasons: the leaves are not reliable characters to differentiate between the species, a number have foliage which is similar; and the leaves of Empetrum are not so loosely arrayed

47

on the shoots, nor do they approach those of Phyllodoce in size. This species is native to western North America from Alaska to California. (Pl. 35.)

P. glanduliflora has a distribution very similar to that of *P. empetriformis* and is to be found high up on the wide open heath-like moors where it is often the dominant plant. It has yellow flowers very similar indeed to those of *P. aleutica*, but if anything is a more robust plant—at least that is how it appears in cultivation. The stems seem stouter and longer and the flower heads carry a greater number of blooms. The ultimate height of each plant may be similar, but the mat-forming habit of *P. glanduliflora* is more pronounced in the early years. An interesting white-flowered form considered to be a natural hybrid between this species and *P. caerulea* and found on Parker Ridge, Alberta, is now available and should be obtained by collectors of phyllodoces. It is described as white, but the shade tends to be light cream rather than of a milky whiteness (Pl. XId).

P. × *intermedia* is a natural hybrid between *P. empetriformis* and *P. glanduliflora*. Growing together as they do, their natural distribution overlapping in many areas, it was inevitable that mixed communities would emerge. Bearing in mind the different colours of the species one could only expect the race to produce a colour range and this too is true. No doubt many names could be applied to the various forms, but two distinct selected clones are *P.* × *intermedia* 'Fred Stoker' (Pl. XIb) and 'Drummondii'. The first is a robust shrublet which ultimately reaches twelve inches in height, and every May completely smothers its foliage with light purplish flowers. These are borne in many-flowered clusters. *P.* × *intermedia* 'Drummondii', on the other hand, has a completely different growth habit, being loose and prostrate. Although the inflorescences are just as amply furnished with flowers they are not so numerous that it is impossible to see the foliage, and the colour is a shade of dark purple.

P. nipponica is a most distinctive species, Japanese in origin as one would expect and bearing pendent, open bells, pure white in colour. The sepals, too, are attractive and, allowing for a little variation within the species, may be bright green or red. This contrasts well with the blooms, and when it is appreciated that the colour of the sepals continues into the pedicel or flower stalk, and this is of translucent texture, allowing the sun's rays to shine right through, it is apparent why most plantsmen grow this species. (Pl. XIa.)

×PHYLLOTHAMNUS is one of those chance plants which cause a stir when they emerge. This one is assumed to be a bigeneric hybrid between Phyllodoce and Rhodothamnus. It appeared unexpectedly in Cunningham and Fraser's nursery in Edinburgh around 1845. This firm grew many ericaceous genera, and after botanical examination its parentage was recorded as *Phyllodoce empetriformis* and *Rhodothamnus chamaecistus*. One is native to western North America, the other to the eastern European Alps.

× *P. erectus* owes its specific name to the nature of its growth. This is stiffly upright and well-established specimens of this dwarf shrub may be up to twelve inches in height. In May the pink flowers appear in terminal clusters. These are quite distinctive, being open and upward facing, midway between the shapes of those of the parents. The foliage is mid green and linear, very similar in appearance to that of Phyllodoce. This unique plant is a first-rate dwarf flowering shrub and one to be considered by all who grow miniature, low growing, woody plants.

RHODODENDRON. The name at least is well known in the Western world and is, of course, both the popular and botanical one for this plant. This is a wonderful genus, ranging as it does over a large part of the world's northern temperate zone, species being found in every northern continent. It does cross the equator though in the Malaysian, Java, Sumatra archipelago, and further species are found at the southern extremity in Australia. There are lots of ericaceous plants in the southern hemisphere—we have met a few in this book —but so far those rhododendrons found south of the equator, and many are colourful and desirable, have not been hardy enough to grow out of doors in this country. Only for this reason are they omitted here. The species centre of Rhododendron lies in the Himalaya, and it is from that vast storehouse of plants that the long tentacles radiate. North, south, east and west they have advanced, modifying and evolving over the centuries. Some further offshoots from the main streams have made their own enclaves and developed their own peculiar characteristics. Because of the myriad climatic and altitudinal patterns, the variations occurring in this genus are not to be wondered at and it should not come as a surprise to learn that trees, large shrubs, smaller shrubs, carpeting plants and even some which continually sucker belong to Rhododendron. Scales of ultimate heights at maturity can be misleading, in addition to being off-putting, as it may take a plant forty years or more to reach full size. Too much stress should not be placed on this as most plants are attractive from an early age.

This genus has been the subject of many books and an infinite number of articles. Enthusiasts there are who have ample accommodation for housing the larger species, but so too among plantsmen cultivating smallish gardens can its admirers be found. National and provincial shows from late winter to early summer have classes purely for rhododendrons. To help in the identification of so large a group of plants the whole has been divided into forty-four series. These do not take into account the height of a plant, the area in which it is found or its degrees of hardiness, but they do bring order of a kind to a very large number of individuals. The concern here, however, is to look at those most suited to the peat garden environment, and despite the restrictions imposed there are still a great many which can come into this category. The matter of choice is difficult in any genus, particularly when one is restricted by space. In the case of Rhododendron, so many wild plants are of decorative garden value, as well as of interest, that no small garden could find room for them all, and so this section of the book can contain only a small fragment of the whole. Not all are readily procurable, but then that is a further challenge to the grower. Furthermore not all flower profusely, but, if that is the plant's normal habit, this is no reason for excluding it. Not all will settle in and grow luxuriantly at the first attempt—some never do—and for what reasons it is not quite understood. But no other genus provides as many architecturally interesting woody plants, shelter and, in some species, flowers. Certain of them are fastigiate, some are loose and spreading, others are mound-forming and maintain a complete ground covering. Still more remain low and squat and, while by far the larger number are evergreen, a few shed their leaves in winter. Flower colour or design can be the reason for growing some species, plant form may be why another group is selected, while leaf shape, texture or colour may be the overriding reason for introducing others. The saving of time may be the reason for planting larger specimens.

There is a surfeit of hybrids, and to mention them could only cause confusion as well as add a long list of names. Rhododendron lovers generally fall into two groups: those who like hybrids and those who prefer the species. In publications of this sort, where cultivars are considered, favourites tend to prejudice the selection, and it is for that reason solely that hybrids are almost ignored.

R. *anthopogon* gives its name to one of the series and has long ovate leaves which are strongly pungent when crushed. The clusters of flowers are small and each white or pink-tinged flower displays a flat face which quickly narrows into a tube. Two feet or thereby is the ultimate height of this upright species and, as a general rule, the plant tends to produce its flowers on top of the bush rather than at the side, which can become scrubby and bare-looking.

R. *baileyi* is named in honour of the plant collector Col. F. M. Bailey, who is also commemorated for his discovery of the "Himalayan Blue Poppy", *Meconopsis baileyi*, now, perhaps unfortunately, known as *Meconopsis betonicifolia*. This Rhododendron, discovered in south Tibet more than sixty years ago, has been much admired, particularly in its better forms. It is generally seen as a thin shrub three feet or more in height with two-inch-long elliptic leaves much studded with scales. In some forms one can get the impression that the plants are partially deciduous. The open flowers are reddish-purple in colour and carried in many-flowered clusters.

R. *beanianum* var. *compactum* is a tidier plant than the species, as the name suggests, and is usually seen as a shrub three feet high and perhaps five feet through. The leaves, which are up to three inches long and an inch or more broad, are quite thick. Their undersides are covered with a dense reddish-brown indumentum. Although sparingly flowered, the trusses of scarlet blooms make this more than a foliage plant.

R. *brachyanthum* is a fairly distinctive plant with hard green leaves two inches in length, the undersides being glaucous. The flowers are pale greenish-yellow.

They are smallish, produced in few-flowered clusters, and hang quite perceptibly. In the Royal Botanic Garden, Edinburgh, some specimens forty years old are still only two feet high by three feet across.

R. *callimorphum*, still known to many as R. *cyclium*, is one of those species with tidy leaves which look almost too neat. These are broadly oval, green on the upper surface and bluish on the reverse side, and appear to form a pattern which gives each leaf maximum light. They are up to three inches long. The flowers appear in clusters, are open bell-shaped, medium pink and quite numerous on semi-mature plants. It can be slow or reasonably fast growing depending on the conditions.

R. *calostrotum* in its best forms is one of the most beautiful rhododendrons. It always stays compact and leafy while the large crimson-purple flowers are in the region of one and a half inches wide. The glaucous colouring on the approximately one-inch-long broadly oval foliage makes it easy to recognize. This species is certainly well up on the list of 'musts', but like all other plants good cultivation and healthy growth are essential for it to have the desired effect. There are first-rate clones segregated from the swarm which are worth searching out. The ultimate size of R. *calostrotum* is three feet, but this is attained only after many years. (Pl. XVa.)

R. *calostrotum* var. *calciphilum* is a smaller version with pink flowers and light foliage.

R. *campylogynum* can present a most attractive picture in May with its numerous, prominent, thimble-like flowers suspended, yet held clear of the foliage, on stiff pedicels. Often only one flower is produced from a flower bud, but two or even three are not uncommon. The colour may be pink, light purple or shades of rose. When allowed to grow without being suppressed or overshadowed it becomes a solid bank of vegetation. It may eventually exceed twelve inches in height, but this may take many years. The three-quarters of an inch long leaves have glaucous undersides. (Pl. 49.)

R. *camtschaticum* is a most interesting species and one that is said not to be easy in cultivation. It's not so difficult, however, provided its shallow roots are not allowed to dry out. Top dressing with peat helps to retain vigour and encourages the stems to root. Old gnarled shoots, with bark similar in colour to *Acer griseum*, up to six inches high after thirty years, does not describe a rampant plant, but layering the shoots helps it to spread more quickly. R. *camtschaticum* is deciduous and has most attractive oval leaves. These are bright green, thin in texture and provide the ideal backing to the reddish-purple open flowers.

R. *canadense* is another deciduous species, but in this case is of fastigiate growth. It is extremely twiggy and at the ends of the shoots it produces clusters of purplish flowers which are sometimes white in the centre. The leaves tend to be linear, and on the upper sides show a dull bloom. This is not one of the finer species, but it displays a different growth habit from that of most other rhododendrons of a similar size. Eventually it may reach two and a half feet, but usually much smaller specimens are seen. Its specific name discloses its country of origin.

R. cephalanthum closely resembles *R. anthopogon*, but differs from it by having leaf bud scales which are persistent. The undersides of the small elliptic leaves are covered in scales, while the upper surfaces are dark green in colour. The narrow tubular flowers which splay sharply at the ends are white or pink or shades between. This shrub will grow in an upright manner and may reach up to four feet in older established specimens. A most desirable variant is *R. cephalanthum* var. *crebreflorum*, for this is a miniature of three or four inches, but unfortunately it is not easily induced to flower. Native to Assam, it can suffer from our harsh late winter weather.

R. chameunum belongs to the series Saluenense which produces thin, flat-faced, outward-pointing corollas. The rich rosy-purple blooms are also decorated with deeper coloured spots. It is one of many collected in Yunnan. The plants, which are slow growing, tend to have flat tops, so that the flowers are always prominently displayed. The ultimate height is around two feet.

R. chrysanthum is allied to *R. ponticum*, but this species barely reaches twelve inches in height. For so small a plant it has large leaves, three inches long and up to half as broad, but they clothe the plant to ground level, making a leafy mound. This is not a plant which flowers freely, in fact for many years it may be flowerless. They are borne in trusses of six to eight blooms which are yellow, as in the usual form of *R. caucasicum*.

R. chryseum has tiny leaves and bears clusters of pale yellow flowers. This makes a change among the alpine Lapponicum types, many of these being shades of purple or blue. All this group make wonderful feature plants as they are hardy, evergreen and can be relied upon to flower annually. Most like an open site, for many of them grow on the exposed upper slopes in Tibet and Yunnan. This is one of the parents of that popular hybrid *R.* 'Chikor', the other being *R. ludlowii*.

R. ciliatum is unfortunately a tender plant, or at least the blooms are tender, for even in the bud stage if the scales have started to open the bells are liable to be damaged by frosts. Nevertheless a spell of fine weather can allow such a glorious display that there could be a danger of overplanting. Well named *ciliatum*, its broadish leaves are bristly with hairs. Six feet is mentioned as its height in the Royal Horticultural Society's Rhododendron Handbook, but this cold statistic does not tell us that the plant can cover itself in pink bell-shaped flowers for the thirty years or more it takes to reach that height. Numerous hybrids are closely associated with this species; two of the best known are *R.* × *cilpinense* and *R.* × *praecox*, the other parents being *R. moupinense* and *R. dauricum*.

R. edgarianum is a hardy Lapponicum type. It is also slow growing with very tiny leaves. Rarely are more than two of its rose-purple flowers produced from one bud, and often only a single flower develops. It is another all-year-round plant, however. Its synonym *R. oresbium* is probably just as familiar to many Rhododendron enthusiasts.

R. fastigiatum varies to some extent, but the flowers appear in a cluster and are so short stalked that the little inflorescence suggests a fairy's posy. The colour, as in most lapponicums, is a shade of purple, this one being lightish. Like heather, this species enjoys an open site where, if free from suppressing neighbours, it will remain furnished with greyish foliage down to ground level. It forms dense hummocks, and after twenty years need be only nine inches high. (Pl. XVb.)

R. *flavidum* is another yellow-flowered species in the Lapponicum series. So many of these small rhododendrons are blue or purple flowered that a variation is always welcome. It was introduced into cultivation from China at the beginning of this century and has proved to be hardy. When crossed with R. *sulfureum* it gave rise to the very popular hybrid R. 'Yellow Hammer'.

R. *fletcheranum* is a recently described species which, until it had been redetermined, had masqueraded under the name of R. *valentinianum*. It appeared to be a frost-resisting form of that species which itself is a most beautiful one, but unfortunately is not reliably hardy in our gardens. This plant, collected by Kingdon Ward in 1932 in south-east Tibet, blooms in March or early April and can be grown successfully out of doors; some raised from the original seed are still in robust health. Like so many of the early flowering species, it is the flowers which are susceptible to frost damage. The foliage, although green, sometimes imparts a brownish sheen, but this is due to the hairs on the leaf surfaces. The large, pale lemon-yellow, open funnel-shaped flowers are most decorative. Plants grown completely in the open remain dwarf and compact, but this can vary according to the amount of shade present.

R. *forrestii*, named after the well-known Scottish plant collector George Forrest, is an extremely variable species. Although some forms flower better than others, none can be reckoned to provide a mass of flowers in spring. The whole plant, never more than a few inches high, is finally a mass of interlacing shoots and these are amply covered with dark green oblong leaves. It is easily distinguished by the purple backs to the leaves. In the seasons when flowers are produced this plant can be quite colourful. The attractive, bell-shaped corollas, sometimes borne in pairs, are bright scarlet. Unfortunately, as the flowering season of R. *forrestii* is March and April, damage from frost is not only possible but likely.

R. *forrestii* var. *repens*, despite the varietal name, is more robust than the species, but it is also shy to flower. That flowers are produced, and first-rate flowers at that, is proved by the hybrid R. 'Elizabeth', which claims this variety as one parent. These hardy evergreen rhododendrons are certainly ideal at providing ground cover. (Pl. 50.)

R. *glaucophyllum* is a good foliage plant which has the most intense blue undersides to its leaves. Turning them over is almost like changing night into day. Established plants usually flower well and the flowers are wide open bell-shaped, rose in colour and carried in many-flowered trusses. A form seen less often is R. *glaucophyllum* var. *luteiflorum*, which is graced with yellow flowers. Eventually this plant may reach four feet, but such a measurement will take a little time.

R. hanceanum, although classified with the Triflorum series of rhododendrons (a group to which the desirable fifteen-foot-high *R. augustinii* also belongs), is of dwarf stature. It forms a low dark hummock, twelve inches in height, made up of greenish-brown oval leaves. Although the clusters of creamy-white flowers are freely borne they are not particularly outstanding. The demand, however, is for the much more dwarf variety *R. hanceanum* 'Nanum'. The leaves in this instance are no smaller, but the blooms in the five- to ten-flowered trusses are a rich shade of yellow. It never exceeds half the size of the true species. (Pl. 51.)

R. hippophaeoides is likened to the "Sea Buckthorn". In some ways this is so, as in the silverish colour of its leaves, but they are typically Rhododendron and are broadly lanceolate in shape. The flowers vary slightly in colour from blue to lilac, but all forms flower well and are hardy. It is very similar in growth to *R. fimbriatum*, but in that species the flowers are rose coloured and the foliage is green on top.

R. hypolepidotum is sometimes classed as a variety of *R. brachyanthum* as it is in the Royal Horticultural Society's Rhododendron Handbook. Like that species, it has greenish-yellow flowers and seems to differ from it mainly in the density of the scales on the undersides of the leaves, which are extremely numerous.

R. impeditum is so neat and perfect in its rounded form that it must be almost one of the first in a list of rhododendrons compiled by the rock gardener. Provided nothing is allowed to influence its natural development, i.e. no part of the plant is overgrown by any other type, either shrubby or herbaceous, its shape will be much admired. Watching a shrub grow and expand is a most satisfying experience, yet to record the growth increment on this species may be difficult in some years. Specimens two feet by a foot can be twenty-five years old. The flowers may appear only singly, but there are so many tiny shoots that in good flowering years no leaves are to be seen. Clones or good forms of most species have been selected in past years, and many have won awards of merit, *R. impeditum* being no exception. (Pl. 52.)

R. imperator is the first of the very dwarf Uniflorum series to be mentioned here, but it must also merit a place in the first six. It is regrettable that this is one of a number of Burmese species which are not completely hardy in all gardens. It was discovered by Kingdon Ward. It is also unfortunate that its flowers are susceptible to the least touch of frost, but where a mild spell precedes its flowering period, which is May, it can be one of the brightest plants in the garden. The flowers are without spots and are pinkish purple, but are so large that they more or less lie on top of the sparse foliage. The lanceolate leaves are dark shiny green and are spaced so as to give the impression that the plant is semi-deciduous. The shoots lie across the surface of the soil, mat-like in effect. (Pl. XIVb.)

R. intricatum, when young, can be tall and spindly, the vigorous young shoots seeming to extend too quickly. These straggly plants, twelve inches high and with only a few flowers, are not so attractive. In subsequent years, however, the branches cease to extend at anything like the initial rate and the sides quickly fill up until the plant becomes well furnished with branches and foliage. The numerous leaves are very tiny and so too are the flowers, but these are gathered together in little bundles so that the plant seems topped by posies of a lilac shade.

R. keiskei is native to the Japanese islands of Honshu, Shikoku and Kyushu, where its habitat is wooded mountain slopes. It is named in honour of the botanist Ito Keisuke. This is a most distinctive species which has quite large, yellow, typically Triflorum type flowers and lanceolate leaves two to three inches long. One outstanding feature is the young foliage which is bronze to browny-red, a shade which is evident for most of the year.

R. keleticum is everyone's plant. It is extremely easy to grow and is also completely hardy. It never looks untidy unless of course some vigorous leafy plant has grown up and suppressed part of it. Being dwarf and compact, it fits into most schemes without dominating or in any way spoiling an arrangement. The leaves themselves are fairly distinctive, the blade coming to a sharp point, while fine hairs decorate the leaf margins. Rarely do the dark green leaves exceed half an inch, but they grow in very close formation. The wide, flat-faced flowers, up to an inch across, are primarily deep purple, but this is relieved by the numerous blood-red spots which bespeckle the throat of the corollas.

R. kongboense, discovered by Kingdon Ward in eastern Tibet in 1925, has been valued as an interesting species from its first introduction. It carries its small rose-coloured blooms in tight terminal trusses. The leaves, less than an inch long, are densely covered on their undersides with a brown felt.

R. kotschyi is in many ways different from any Rhododendron so far mentioned. It belongs to the Ferrugineum series, which bears the name of the much admired "Alpen Rose". It could be described as a miniature of that species, for it has the same sort of inflorescence, with long, tubular, rose or sometimes white flowers, but smaller, opening out in a trumpet-like flare, a description which could be applied to *R. ferrugineum*. The undersides of the leaves, too, are covered by numerous brown scales. Plants of eighteen inches and more can have their age reckoned in as many years.

R. lepidostylum is one of the most attractive foliage plants in the genus. For most of the year its leaves are blue, although naturally in winter this colouring is less intense, and while the foliage is young there is no more conspicuous plant. This is a very compact shrub with ovate leaves a little more than an inch long, overlapping, so that from an early age the soil is completely hidden. The fact that the leaves are bristly with stiff hairs is an added feature. May is the flowering month, but even on old specimens flowers are not plentiful and are often hidden among the foliage. While the pale yellow corollas blend ideally with the greenish-blue foliage, and are in themselves of high value, their shortage is not apparent in a shrub renowned for the quality of its leaves. (Pl. 53.)

R. lepidotum is a distinctive plant, although there are many variations. It has scattered, densely scaly leaves which are only an inch or thereby in length and are so sparse that the shrub appears semi-deciduous. (Pl. 54.) This is particularly true of *R. lepidotum* var. *elaeagnoides* which has pale yellow nodding flowers, appearing when the plants are only inches high.

R. leucaspis is one of the most ornamental dwarf rhododendrons growing in our gardens. It is a great pity that one must question its hardiness, for it has proved tender in inland gardens; its flowering season, too, is outwith the ideal months for blooming, being March and April or, in certain years, even February. In a slightly sheltered site, shaded from early morning sun, the blooms can be encouraged to open and it is then that this species is most admired. The pure white flowers in twos and threes are wide open and may measure two inches across, but it is the contrast in the well-spaced chocolate-coloured stamens that complements petals and stamens so admirably. The foliage itself is of an attractive texture and colour, being dark green and covered with fine down-like hairs. At certain times the pubescence is golden. Even young plants flower well.

R. linearifolium is most unusual for here, as the specific name suggests, the leaves are exceedingly narrow. They are also very hairy, which overall gives the shrub an 'unrhododendron-like' appearance. The foliage is deciduous and is topped in May by the rose-pink flowers, the petals of which are divided into linear strips. It is of Japanese origin, although the validity of the name is doubtful.

R. lowndesii is a difficult plant to grow and so far its general needs are not quite understood. When they have been properly assessed and the species is as plentiful as *R. pemakoense*, it will be considered an essential plant for any worth-while collection of rhododendrons. It was discovered as recently as 1950 by Col. Donald G. Lowndes in Nepal. In gardens, and mostly in pots or pans, this species attains only a few inches in height. It is extremely twiggy, with no really thick stems. The bright green deciduous leaves are bristly with hairs, while the large flowers, up to one inch in diameter, are translucent yellow, shading to light pink in the centre. This is a species that rewards patience and will surely become a peat garden plant when more is known about its cultivation.

R. ludlowii is another dwarf spreading shrub with large flowers. These are very pronounced, sitting well clear of the small, ovate, scaly foliage. The colour of the cup-like flowers is basically yellow, although this can be much marked by pink-shaded areas and a dense cluster of brownish-purple spots on the upper part of the flower. In general *R. ludlowii* tends to radiate from a thick stem, disposing its leaves sparingly on the branches so that the plants are inclined to look very open, and in some instances, although a true evergreen, partly deciduous. Despite the fact that the late Frank Ludlow discovered this species in 1936 it has never been plentiful. It is certainly extremely slow growing and it can be difficult to establish in some areas and this no doubt has discouraged many would-be cultivators from making the attempt to grow it.

R. megeratum belongs to the same series as *R. leucaspis*, and like that plant is a little tender. Its blue-green foliage is accentuated by the bright glaucous colouring

on the undersides. The leaves and flowers are not so large as those found in
R. *leucaspis*, while the flowers are yellow, not white, but the stamens have the
same distinctive chocolate shade. A well-flowered specimen is an asset to any
group of shrubs.

R. *microleucum*, a name not listed by many books, is a most attractive dwarf plant
of the Lapponicum series. It has tiny leaves and finally grows into a most
compact shrub not unlike R. *scintillans*, except that it has pure white flowers.
(*See* Pl. Ib.)

R. *moupinense* flowers in February and March. At such a time of year flowers are
most welcome, but it also means that the blooms are frequently damaged by
frost. Therefore it is not a plant for the most exposed part of the garden if one
wants to enjoy the blooms, although, apart from flowers, it is quite hardy. A
site in the background among the taller shrubs would be the most likely place
for it to succeed. The flowers, which are open bell-shaped and up to two inches
across, may be creamy white, light or deeper pink. It has been used as a parent
when making crosses, and the one most widely known is the floriferous R. ×
cilpinense.

R. *myrtilloides* is native to Burma whence it was introduced into British gardens
around 1914. It is so very closely allied to R. *campylogynum* that quite often it is
treated as a variety of that species, but it has smaller flowers and is in fact a
smaller plant. The wine-coloured, dainty, bell-shaped blooms are waxy, but
quite translucent when lit by the sun from behind. The leaves too are tiny
and are often reddish-brown in colour. It is completely hardy and will grow in
most exposed places, provided it receives sufficient moisture at its roots.

R. *nitens* must be included in any collection of rhododendrons if only because of
its time of flowering, which extends into July. One of the last of the dwarfs to
bloom, it has wide-open flowers which are rich reddish-purple amply spotted in
parts with deeper coloured dots. The flowers are held well clear of the small
evergreen leaves. It may finally reach fifteen inches in height, but after as many
years it could still measure less than a foot.

R. *nivale* is another of the Lapponicums so valuable to the scale of small gardens.
It is a rare plant, however, and considered by some to be difficult to grow.
More accurately, perhaps, it could be said to be difficult to propagate. It is
found high in south Tibet, Nepal and Bhutan, and although the name *nivale*
means snowy it is because of its habitat that it has been so named, and not the
colour of its flowers, for these are purple to violet. They are produced singly, up
to half an inch long, and appear in May when they sit on top of the small,
quarter-inch-long scaly leaves. The rate of growth can be as little as a quarter of
an inch per year. Plants in cultivation today are from a collection made by R. E.
Cooper, at one time Curator of the Royal Botanic Garden, Edinburgh.

R. *obtusum* is an extremely floriferous Japanese species and well worth growing, but
it is not recommended for the peat garden. Its colours are too harsh and tend to
clash with the other less flamboyant plants. One could almost say that by

leaving out *R. obtusum* the peat garden would be more colourful and certainly it would possess a better quality.

R. patulum belongs to the Uniflorum series, but the bells are so large when compared with the size of the foliage that the thought of one flower per shoot need not put one off growing this species. A plant in isolation makes a solid mound of foliage (which in itself is an attractive feature), but in April when the plant is in bloom leaves need not be visible. The lilac-purple drooping flowers appear so heavy that they give the impression of being suspended on thin pedicels. Like *R. pemakoense*, the next species on our list, the blooms are frost tender and it can happen that the promise of a wonderful display may be ruined overnight. (Pl. XIVc.)

R. pemakoense is inclined to spread by suckering shoots when happy in its situation. Young plants placed straight into the peat blocks have this tendency, and it is a trait which is to the gardener's advantage. When only a few inches high it can produce a great number of open funnel-shaped bells. Obviously it would be a mistake to plant this species anywhere except in the front as the effect of its branching system which clothes the plant to the ground would be lost if taller plants obscured it in any way. *R. pemakoense* is native to Tibet. (*See* Pl. Ib.)

R. primuliflorum is well named, for the tight clusters of flowers resemble the arrangement and shape of certain types of Primula. It belongs to the series Anthopogon in which the corolla is made up of a long tube almost closed at its throat, yet the petal lobes flatten out at right angles to the axis of the white to rose flowers. The leaves may vary between oblong and oval. Fully grown plants four feet high are commonplace, but those attaining a height of only two feet are just as decorative.

R. prostratum has small glaucous green foliage, and as the growing habit of this shrub is low and flat the plant can, at times, look like a blue mat. Top dressing in early spring will prevent the twiggy shoots from becoming hard and woody as they are inclined to project into the air slightly. This encourages it to remain vigorous and the foliage to retain its normal size. The flowers sit proudly on top of this small carpet, being held upright on erect stalks. The wide-open corollas are rose-crimson with deeper coloured spots bedecking the upper petals.

R. proteoides is a rarity. It is included because it is also a garden-worthy plant. The name means like a Protea, that South African genus with closely packed flowers carried in a head almost immersed in the upper leaves. The flowers of this species are carried in a compact inflorescence and are creamy white to yellow, at times flushed with pink, but this floral description is almost academic as *R. proteoides* is shy to bloom. Once more it must be said that this is almost immaterial as one of the decorative features of the plant is the thick woolly pile on the undersides of its oblong leaves. A feature of the leaf is that it reflexes its edges so that the reddish brown felt is readily visible. All books say this shrub is slow growing, but to put meaning into this statement a specimen thirty years old may be a mere nine inches in height and be less than eighteen inches across. It belongs to the Roxieanum series with natural habitats in Tibet, Szechuan and Yunnan.

R. pumilum can reach twelve inches in height, but mostly it is seen as a low spreading plant of a few inches and grows wild on the Burma-Yunnan border on open slopes. It must form part of the heath in these areas and no doubt is kept dwarf by the intense light conditions. Because it is shallow rooted its gravest danger in cultivation is from drying out, and the cool environment of the peat garden seems to be able to provide conditions similar to those in which it is found in nature. It is really a dainty shrub with small thimble-like buds for flowers. These are pink in colour and little more than half an inch long. (Pl. XIVa.)

R. racemosum is one of the hardiest of the dwarf species. In peat or rock garden where this plant flourishes self-sown seedlings are commonplace, and in sheltered and exposed situations it will flower prodigiously despite growing in an almost soilless crevice in full sun. Healthy, well-furnished plants, however, need a more favourable site, and when this is provided *R. racemosum* becomes a most desirable species. It has dryish grey-green foliage, although the undersides of the leaves are almost pure white. The range of flower colour is white to deep rose, and one of the best, which is dwarf and compact, is a form collected by George Forrest and carrying the number Forrest 19404.

R. radicans is mat-forming and completely covers the soil with a two- to four-inch layer of small oblong leaves. It dare not grow taller in nature since its habitat is open stony moorland in Tibet. This hardy, dark-leaved evergreen produces its flowers irregularly. They are of the flat-faced type up to an inch or thereby across, purple in colour and produced singly on stems which just keep the blooms clear of the foliage. It is most effective when planted on top of a solid peat block.

R. radinum has small leaves for a plant which will eventually reach four feet in height. It will also spread to the same extent, but from a bought plant of, say, six inches this may take forty years. It is densely furnished with hundreds of narrow leaves less than one inch long and a quarter of an inch wide, and without doubt the best specimens are found where light is not restricted. The smallish flowers are borne in tight clusters, posy-like, and are rosy pink. In full bloom or devoid of flower *R. radinum* is a plant of character.

R. roxieanum is most distinctive and much in demand by Rhododendron enthusiasts. It is extremely slow growing and in consequence slow to reach a flowering state, but even without flowers it is a most desirable foliage plant. The long narrow leaves, up to four inches long by three-quarters of an inch wide, make this a prominent shrub, the leaves being dark glossy green, but densely covered underneath with brown indumentum. When flower trusses appear, up to fifteen white to deep pink blooms may be carried in each truss, but because of its reluctance to flower the colour is not valued in the same way as in other species; nevertheless none is a poor plant.

R. russatum may still be known to many as *R. cantabile*, but under whichever name it is grown it will always be classed as first rate. It has a neat rounded habit with well-balanced leaves. These are approximately one inch long by half an inch wide and are dark green yet scaly on top and a russet shade beneath. For a member of the Lapponicum series the flowers are quite large, being almost three-quarters of an inch long and widely funnel-shaped. The colour varies between deep purple and violet and five or six flowers make up each truss. Its habitat is in Yunnan, from where it was introduced by George Forrest around 1918.

R. *saluenense* varies a great deal in its growth habit, some forms being semi-prostrate while others have a more upright stature. The flowers occur in pairs or in threes and have short tubes, but they open widely where the petals separate. The rich rosy-purple shade of the flowers makes this a most attractive species, the upper petals being dotted with spots of a deeper hue.

R. *sargentianum* is a dwarf, tight-growing shrub named after C. S. Sargent, at one time Director of the Arnold Arboretum. It has clusters of small tubular flowers, typically Anthopogon series in shape, and these may be white, creamy-white or pale yellow. This plant was first discovered by Ernest Wilson in 1903, and one of the best forms with clear yellow flowers was collected by him in Mupin, western Szechnan, and is identified under the number W 1208. It bears tiny, elliptical, shiny green leaves which are strongly aromatic when bruised, a feature not uncommon among these dwarf Himalayan alpine rhododendrons.

R. *scintillans* must be one of the first dwarf species on any list. In its better forms it has flowers which are almost royal blue and there is a particularly fine clone, bearing lavender blue flowers. It is a very twiggy plant which eventually forms a solid dome of tiny, bronzy-green leaves, the colour, due to its flowers, changing to shades of blue or purple each May. Young plants can be just as satisfying, and their sprawling individually vigorous shoots give the plant a more informal appearance. It is on less mature plants that the best flower trusses are formed.

R. *serpyllifolium*, since it bears the smallest leaves in the genus, must have a place in the collection. These tiny deciduous leaves are between a quarter and half an inch long, and are thyme-like in appearance. The small pale pink flowers are usually in pairs. Although the leaf measurements may be classed as microscopic the plant itself may eventually reach four feet. It belongs to the Azalea series and, as though by dint of its tiny, perfect proportions, is native to Japan.

R. *trichostomum* is very similar to R. *radinum*, in fact the latter is often considered to be a variety of it. It differs mainly from R. *radinum* by having a corolla relatively free from scales. Obviously, unless one were deeply interested botanically in the genus, one of these rhododendrons would suffice as they are very similar. (Pl. 55.)

R. *tsangpoense* has leaves which are broadly oval. In some ways the foliage resembles R. *racemosum*, being grey-green on top and glaucous beneath, but this is a taller shrub with a much more open arrangement of branches. Its flowers, however, are completely different, being in small groups with a maximum of six bell-shaped blooms, pink to violet in colour. Some have a grape-like bloom. While the species is ideally suited to the background, the form most appropriate for the front terraces is *R. tsangpoense* var. *curvistylum*. In this the flowers are a shade of pink.

R. williamsianum was introduced into our gardens early this century, having been discovered by Wilson in 1908 in eastern China. It honours Mr J. C. Williams, also of Camellia fame, who helped to finance many of the professional plant hunters in the early part of this century. This species is one of the neatest known to man and in every way is a beautifully balanced plant. Completely hardy, evergreen and virtually free from plant troubles, it decorates the garden all year with its roundish leaves and compact hemispherical habit. As a rule young plants do not flower freely, but once a certain stage is reached no finer sight can be seen than this species smothered under its large, perfectly fashioned, delicate pink bells. (Pl. XVc.)

Naturally the hybridists have used *R. williamsianum* as a parent when making crosses, for few plants have as many desirable features to pass on.

R. yakusimanum is the final Rhododendron to be mentioned in this work, and it can be said that few genera have a more worth-while species on which to hang up the spade. This is a dwarf species often seen grown in large pots for display at flower shows. On these occasions plants a foot high and two feet across carry as many as fifty trusses of flowers. These are pink at first, but, as the flowers age, the colour fades until the blooms are almost pure white. In the garden the plants tend to grow a little more strongly and perhaps flower less profusely. As a point of interest, in a recent Japanese flora the spelling of the specific name is changed to *yakushimanum*, as the island from which it gets its name is *Yakushima*. (Pl. 56.)

RHODOTHAMNUS is a monotypic genus from the eastern Alps and the Dolomites. In nature it is found in limestone areas, but in cultivation the inclusion of lime in the soil is not necessary. Reginald Farrer extols the virtues of this miniature shrub and says it is easy, although he admits to its having transplanting problems. It is not all that easy.

R. chamaecistus is a beautiful little plant which can be encouraged to grow in a well-drained peaty soil. Top dressing with peat in spring helps to keep the fine surface roots moist and cool, although planting in a not too exposed area will also help to keep the temperature reasonable. Its lack of stature demands that it be allotted a place at the front of the peat garden where it can be watched and admired when in flower. The evergreen leaves are tiny and glandular, while the open flowers, appearing in twos and threes at the tips of the shoots, are pale pink in colour. At one time it was considered to be a Rhododendron and was so described in the *Botanical Magazine*, plate 488, in 1796.

TRIPETALEIA is an unusual plant to find in this family. As its name suggests it has three petals, normally a number that is more associated with monocotyledons. There are two deciduous species involved and, like so many other less common plants, they are native to Japan.

T. bracteata forms a mass of stout stems and, although it is said to grow to six feet, specimens thirty years old are still only eighteen inches high and flower annually. The petals are greenish-white flushed with pink and are curiously twisted at their tips.

T. paniculata in many ways resembles the previous species. Here, however, although bearing smaller blooms, these are carried in large panicles. It is also recognizable by its joined sepals. Both are completely hardy and in nature inhabit the high mountains of Hokkaido and Honshu. It is August before the flowers are seen.

TSUSIOPHYLLUM is not perhaps amongst the first shrubs one would think of planting, but is an interesting one that collectors are bound to seek out. It is partially deciduous, this depending on the severity of the weather, is slow to grow, specimens of thirty to forty summers being no more than a foot high, and is compact in its growth.

T. tanakae is the sole member of the genus. Its distribution is given as the mountains of Honshu, where it blooms in July. It is said to be rare in nature or at least not plentiful. The tiny white flowers are also sparse, being produced singly or in twos or threes so that the plant has a white-studded appearance rather than one covered in bloom. It is closely akin to Rhododendron; in fact a recent Japanese flora lists it as *Rhododendron tsusiophyllum*.

VACCINIUM was at one time considered to be in its own family, Vacciniaceae, because of its inferior ovary, but the most up-to-date floras still include it in the Heath Family. It is a large genus containing upwards of 130 species which are naturally distributed over a large part of the cooler northern hemisphere. It also occurs in tropical climates, but in those latitudes its habitat is invariably in the high mountains. Most of the species are perfectly hardy in Britain, the few exceptions being evergreen from lower altitudes and native to the southern part of North America, Ecuador and farther afield in China. Even they can be grown successfully out of doors in favoured gardens. Vacciniums include the well-known "Cranberry", "Bilberry", "Whortleberry" and "Blueberry".

V. angustifolium was at one time listed as a small form of *V. pensylvanicum*, a name no longer considered valid. This low-growing deciduous species, native to the north-eastern parts of North America, increases by suckering. It does not do this rapidly, so is not a menace. Old stems, however, tend to become woody and twiggy and regular pruning and top dressing will encourage strong shoots. This is important as it is the autumn colour of the leaves and strong young shoots on which the gardener relies for colour. Its flowers are typically bell-shaped, white flushed with pink, and these can be followed by blue fruits. This is the "Blueberry" of the United States where large areas are cultivated.

V. arctostaphylos forms an open bush of five feet or more but it need not exceed the shorter height if thinned. It can of course grow much taller but it deserves to be considered because of its autumn tints. The leaves of this Caucasian species are approximately three inches long by one and a half inches wide, and they can change from green to deepest purple before being shed. It is an extremely useful plant to provide a certain amount of shade in summer, yet in winter, as it is deciduous, it impedes light very little. Slender climbers, e.g. some of the Codonopsis, can be trained to scramble through the branches without doing harm to the host.

V. caespitosum is the "Dwarf Bilberry" of the North American continent. It usually grows to six inches in our gardens and colonizes an area most effectively. It is not invasive to the extent that it soon suppresses its neighbours, but its progress should be watched if long-lived, dwarf shrubby species are growing near by. Its autumn colours demand that it be included as its deciduous leaves change to red and purple before falling. Throughout the winter months its red twiggy shoots are themselves an asset.

V. corymbosum is of upright growth, not quite so tall as *V. arctostaphylos* and producing more twiggy basal shoots and suckers than that species. It serves the same purpose, for it is just as colourful in autumn and can be considered as an alternative. It is the "Tall Blueberry" and is worth while in its own right. Unfortunately these North American species do not appear to produce their fruits in quantity in this country. There is a variety sometimes seen, *V. corymbosum* var. *amoenum*, which seems more dwarf in growth, is just as colourful in autumn and certainly flowers well. It is probably of hybrid origin.

V. delavayi is most distinctive and, as the specific name suggests, commemorates Jean M. Delavay, a French Abbé who collected botanical specimens in west China. It is not a plant grown for its flowers, as these are rarely seen, but it has neat round foliage which can be bronze in colour when young before changing to green. Plants two feet high are quite old and normally they then measure three feet through. The growth is compact and the plant evergreen, so that all light is excluded from the soil. Occasionally long shoots develop which tend to be densely hairy, but these should be left unpruned, for during the next growing season numerous side growths develop on their upper half.

V. glauco-album was first discovered in Bhutan as long ago as 1838, although its introduction into our gardens is probably much later than this. It is a handsome species with oval leathery leaves which are outstandingly glaucous on the upper surfaces. These leaves are approximately two and a half inches long and are borne alternately on long arching shoots which often arise as sucker growths some distance from the older stems. It is on the more mature branches in the axils of the leaves that the slightly pink-tinged close-mouthed bells occur. Later, in autumn, the blue-black fruits appear, at first being densely covered with a most attractive grape-like bloom.

V. macrocarpon, sometimes listed under the generic name of Oxycoccus, is the

"Larger American Cranberry". It has thin wiry shoots which radiate from the established centre of the plant. These may reach twelve inches or more and be clothed with thinly distributed narrow evergreen leaves. This foliage is glaucous on the undersides. The small pink flowers, the petals of which are separated to the base and curved back to reveal the stamen cluster, develop a short distance back from the tips. The large blue-purple fruits should, in normal circumstances, follow in autumn. It is an interesting plant for both foliage and stem tracery.

V. mortinia, a species reputed to be tender, refutes this statement at Edinburgh. It is native to the Andes of Ecuador, where it grows at an elevation of around 11,000 feet. In the garden it grows into a shrubby mound two feet high, is compact and clothed in masses of small evergreen foliage. The term evergreen really refers to the older leathery foliage, for the young leaves and tips of the shoots are pink in colour. Masses of flowers develop underneath the arching shoots and these are followed by dark purple fruits. The plant, especially when young, is extremely hairy, giving the freshly developed growth a matt finish.

V. moupinense is a small shrubby plant of twelve inches or thereby and owes its attractiveness to its evergreen leaves and their arrangement. It spreads slowly, but is so densely leafy that it forms a solid hump of vegetation. The foliage is much the same as described for *V. delavayi*, being two toned, for the young leaves are golden brown, but change to green as they mature. Planted near the front of the peat garden this species, apart from being decorative, is a shelter for some low-growing species on its leeward side.

V. myrtillus is the "Blaeberry" or "Bilberry", a fairly common British native species on heaths and moors. It increases naturally by underground runners which sucker freely, making a thicket of twiggy shoots. The winged stems are often green and in summer carry the small ovate dull green leaves. The fruit is black and appears in early autumn, following the fertilization of the pink flowers, which open in April. A form on which white fruits occur is known as either *V. myrtillus* var. *leucocarpum* or var. *album*.

V. nummularia, recorded as epiphytic in nature, is one of the many plants admired by alpine enthusiasts because of its ability to flourish in a pot. Again it is one which is slandered by the phrase 'not hardy except in south-west gardens'. At Edinburgh in a north pocket in the rock garden there is a plant at least forty years old, and growing in peat blocks in the peat garden there are others which have been there for at least twenty years. This is an extremely handsome evergreen shrub from north China, the stems of which are densely covered with bristles. The round leaves are convex and display quite clearly the inset veins. The leaf margins, particularly near the petiole, are adorned with long ciliate hairs. It is one of Ericaceae's most attractive shrubs. It spreads by suckering, and while never becoming twiggy it does develop two or three side branches on older shoots. May is when the pink-tipped flowers appear and these are carried in terminal drooping racemes.

V. ovatum is a woody plant up to five feet or more in height with most attractive colourful foliage. The ovate leaves, although basically green, are in their early life a reddish-bronze. This is an evergreen, and suspended below the arching shoots numerous white-tinged pink flowers appear in spring. This is a North American west coast species which is quite hardy, although in extra severe weather the tips may be harmed.

V. oxycoccus, the "Cranberry", is a British native species which can be found in many counties, but its distribution is wide, for it embraces the northern hemisphere. This Vaccinium has a creeping habit and sends its thin wiry shoots through and among other peat-loving plants. Its tiny leaves are well spaced on the branches, while in summer, at their tips and on longish stems, the tiny pink flowers develop. The edible but acid fruits follow in autumn.

V. retusum has most attractive foliage when young, for the leaves are covered with a partly glaucous sheen. It is native to India, Himalaya and Bhutan. It is a compact, bushy plant of approximately twelve inches in height and produces clusters of pink bell-like flowers towards the ends of the shoots. Later the purple berries add to the decoration of this evergreen.

V. smallii from Japan, flowering in May, makes a compact much-branched shrub of barely three feet. Its stiffly erect naked shoots are seen when the plant is leafless during winter, but by early summer these are hidden beneath the two-and-a-half- to three-inch-long elliptic leaves. In autumn, when the colour of the foliage changes, the leaves are streaked with red and yellowish-green.

V. vitis-idaea, the "Cowberry", occurs in many parts of Britain. It is often to be seen in woodland where its creeping colonizing stems have a free run through the moist light covering of the forest floor. It finds the compost in peat gardens a most suitable medium and one must keep a look-out for its spreading into other dwarf, slow-growing species. It does make a good ground cover though, for its numerous stems completely carpet the soil. As the flowers are pink bells, borne in short terminal racemes, they are themselves quite attractive, but when the reddish-orange fruit develops, its appeal is even greater. There is a dwarf variety, *V. vitis-idaea* var. *minus*, which is a true miniature, and while its growing habit is similar to that of the species its rate of expansion is very much less.

8 THE PRIMULA FAMILY (*Primulaceae*)

CORTUSA is shade loving and in nature grows in woodland. It is a small genus of one or two species, in appearance is very like a Primula, and its perennial habit commends it to the gardener. It gets its name from Jacobi Antonio Cortusi, who in the latter part of the sixteenth century was director of the botanic garden at Padua, one of the oldest in Europe.

C. *matthioli* carries its rose to purple flowers in an uneven cluster on top of slim eight-inch-high stems. Each bloom, suspended on a curving wire-like pedicel, is bell-shaped and small for the size of the plant. The flowering season is late, July in fact, but it appears to last a long time. The plant makes good clumps which may be divided in spring. The leaves are pale green and crenate, and there is a section of Primula with leaves so similarly shaped that it is called the Cortusoides section.

CYCLAMEN are not plants that one would normally suggest planting in a peat garden, but there is one species which is not difficult to grow and, because it flowers from August to October, is very welcome. Corms are often offered, but they are usually dried-up corky lumps which have been collected in the wild and most people find them reluctant to re-establish. Cyclamen can be raised from seed, but it is possible to buy young growing corms and these are to be preferred. One must make sure the corms are planted the correct way round, however, for in some species the corms root from the top beside the flower buds, so there is just the danger that they may be planted the wrong way up.

C. *neapolitanum* is from southern Europe, yet is completely hardy. Its first flowers appear in August before the beautifully white-mottled foliage appears. The petioles travel underground for some distance before appearing above soil level, and as they emerge, erratically, and the leaf blades expand, the planted area is provided with a solid covering. White, pink and shades of carmine are the colours found in the various forms.

DODECATHEON is the "American Cowslip" and is also referred to as

"Shooting Star". This is a very interesting group of plants native to North America which gets its second common name from the tight yet prominent arrangements of its styles and stamens. These protrude well beyond the sharply reflexed petals, the whole being reminiscent of an arrow head. Although in a small way Dodecatheon is quite spectacular, it is a mistake to overplant it because the foliage withers early in the season, leaving bare patches of soil. Planting through other non-invasive genera such as gentians can be successful. Dodecatheons have the ability to propagate from roots, and clusters of young plants can sometimes be seen sprouting from an individual plant which has lost its crown; they are also easily raised from seed.

D. *clevelandii* is variable in that there are a number of forms included in it. Basically the colour of the corolla tube is dark maroon with a yellow band where the free section of the petals bends backwards. These lobes may vary from magenta to white.

D. *jeffreyi* is one of the taller species. It also has large pale green leaves. The flowering stem can be up to two feet in height and the inflorescence can be many-flowered. The flowers themselves are rich purple and are further decorated with a yellow ring.

D. *meadia* is one of the most popular. It too may reach two feet and carry numerous flowers. These are rose-coloured with a light, almost white, base. There are a number of forms, one of which is D. *meadia* var. *integrifolium*, and sometimes given specific rank. Although not so vigorous, being but half the size of the species, it is still most attractive. Division in March is a ready method of propagation, although seeds can produce plants in quantity in time.

OMPHALOGRAMMA seems rather an unusual plant to be classed in this family, yet it is so closely related to Primula that occasionally it has even been included in that genus. In the first place it has zygomorphic flowers and in some ways could be likened to Gloxinia. Admittedly some primulas have this characteristic, the most noteworthy probably being P. *sherriffiae*. Add to this the fact that Omphalogramma has six petals, and one is surely dealing with factors unusual for Primulaceae. However, the interest they arouse and their decorative value are not in question. They are so distinctive that even the casual visitor wants to know more about them. Unfortunately they are not the easiest of plants, and their cultivation is not generally understood, but an open part of the peat garden provides the best environment. Sometimes they are shy to bloom, but if the clumps are split up and the divisions replanted in a fresh site this often induces flowering. Very little foliage is visible at flowering time in some species, while in others it may be fully developed. Those which bloom before the leaves produce large flowers, usually on short thick stems which hold the large flowers little more than clear of the ground. After flowering the stem lengthens and the seed capsule, still showing its prominent style, by then may be twelve inches above the soil.

O. elegans grows in peaty bogs in Tibet. This is one of the easier species with enormous flowers. These are bluish-purple and have a white hairy throat. The stamens, too, are attractive as they form a cluster in the centre of the flower which appears before the leaves. (Pl. VIIb.)

O. elwesianum has violet-blue flowers not as large as the preceding species, nor is the flower stalk quite so tall, finally being scarcely six inches in height.

O. minus, as the name implies, is one of the smaller species rarely exceeding four inches. It is an extremely hairy plant with reddish-purple flowers and has the tendency (as they all do) to hold its flowers at right angles to the stem rather rather than let them droop. It is native to Burma, Tibet and Yunnan, and has been known for almost fifty years.

O. souliei is often confused with *O. elegans,* and it is doubtful whether the true plant is still in cultivation, although plants bearing that name do exist. When compared, however, *O. elegans* is much more hairy than *O. souliei* and has bigger flowers. Neither flowers profusely even under ideal conditions. Some Perthshire gardens, where the annual rainfall exceeds forty inches, seem to have simulated conditions close to those in nature.

O. vinciflorum is the easy one, if that word can be used when describing omphalogrammas; perhaps it would be more correct to say that it is the one that flowers best. It is also amongst the tallest growing, although some forms appear to flower when only a few inches high. The plant is well provided with hairy, almost clammy, foliage by the time the violet flowers appear. Being completely herbaceous, this species dies back in winter to a distinct bud, the scales of which are found to be orange or beige if they are inadvertently uncovered.

PRIMULA is a most outstanding genus full of desirable and decorative plants which can be enjoyed singly or in drifts. Obviously the size of the overall planting area will determine the quantities, but if small groups irregularly arranged can be included they will do a great deal for effect. We are lucky, too, in that a great many species of Primula are hardy in Britain. If Rhododendron can be credited with supplying the bulk of woody plants for the peat garden, surely no one would object if the honour of providing most herbaceous-type ground cover went to Primula.

In the main primulas are confined mostly to the northern hemisphere, and certainly those with which we are concerned here are from north of the equator. When one thinks that the number of species identified is in excess of 500 (in addition to these there are at least 1,500 synonyms) one must realize that growing space can be found for only a very small selection. How does one set about making a choice? Many of them are so decorative and desirable that the tendency, inevitably, must be to include too many. They grow under all manner of conditions. Some are saxatile and so are less suited for use in this environment. Many are woodlanders, not a few are meadow plants (albeit the high alpine pastures of the Himalaya), although a few European members, the

Ia Peat bank at Braewick, Shetland, showing clearly the root penetration of the moorland/heathland flora.

Ib A corner of the Peat Garden, Royal Botanic Garden, Edinburgh. Prominent in bloom are *Phyllodoce caerulea* (page 41), *Rhododendron pemakoense* (page 58), and *R. microleucum* (page 57).

Ic The Peat Garden in May, the colour here being provided almost entirely by members of the Ericaceae.

IIa *Arcterica nana* (page 30).

IIb *Cassiope selaginoides* (page 32).

IIc *Corydalis cheilanthifolia* (page 113).

IId *Disporum smithii* (page 84).

IIe *Epigaea gaultheroides* (page 35). Photograph:
H. Esslemont

IIf *Fritillaria meleagris* (page 86).

III *Calanthe alpina* (page 108).

IVa *Gentiana* 'Inverleith' (page 125).

IVb *Hacquetia* *epipact* (page 126).

IVc *Harrimanella stelleria* (page 41).

Va *Hydrastis canadensis*
(page 126).

Vb *Kalmia angustifolia*
(page 42).

Vc *Kalmiopsis leachiana*
'Marcel le Piniec'
(page 43).

VIa *Lilium grayi* (page 89).

VIb *Lilium japonicum* (page 89).

VIc *Lilium oxypetalum* (page 90).

VIIa *Orchis elata* (*Dactylorhiza elata*) (page 133).

VIIb *Omphalogramma elegans* (page 68).

VIIc *Paris polyphylla* (page 95).

VIIIa *Meconopsis grandis* (page 130).

VIIIb *Meconopsis integrifolia* (page 131).

VIIIc *Meconopsis chelidonifolia* (page 130).

IX *Meconopsis horridula* (page 13

Xa *Nomocharis mairei* (page 93).

Xb *Nomocharis pardanthina* (page 94).

Xc *Below left: Nomocharis farreri* (page 94).

Xd *Below right: Nomocharis saluenensis* (page 94).

XIa *Phyllodoce nipponica* (page 48).

XIb *Phyllodoce × intermedia* 'Fred Stoker' (page 48).

XIc *Phyllodoce caerulea* var. *japonica* (page 47).

XId *Phyllodoce glanduliflora* (white form) (page 48).

XIIa *Primula polyneura* (page 76).

XIIb *Primula cockburniana* (page 72).

XIIc *Primula whitei* (page 80).

XIId *Primula aureata* (page 70).

XIIe *Primula reidii* var. *williamsii* (page 77).

XIIf *Primula sieboldii* (page 78).

XIIIa *Shortia soldanelloides* var. *magna* (page 147).

XIIIb *Sanguinaria canadensis* 'Flore Pleno' (page 145).

XIVa *Rhododendron pumilum* (page 59).

XIVb *Rhododendron imperator* (page 54).

XIVc *Rhododendron patulum* (page 58).

XVa *Rhododendron calo-strotum* (page 51).

XVb *Rhododendron fasti-giatum* (page 52).

XVc *Rhododendron wil-liamsianum* (page 61).

XVIa *Trillium grandi-florum* 'Roseum' (page 100).

XVIb *Trillium erectum* var. *albiflorum* (page 100).

XVIc *Trillium chloro-petalum* (page 100).

Vernales section in particular, also fall into that category. Others, collected at too low an altitude or in warmer countries, are just not hardy enough to withstand our winters. In nature some are to be found in areas which are very wet, particularly in spring, and if this amount of water is necessary it is very difficult to provide these conditions in smaller gardens.

Most primulas are perennial and many are easily raised from seed. Where large numbers are wanted quickly this is the obvious way, but if a moderate increase will suffice, division of the old clumps can provide a few extra plants.

To facilitate the study of so large a genus some sort of key had to be devised which would classify the species into manageable groups. These groups were named sections and they number around thirty. Two of the most popular are the Vernales section, to which the ubiquitous P. 'Wanda' and its associate hybrids belong, and the Candelabra section, which includes the deep red P. *pulverulenta*. As plant comparisons are a large part of the interest in growing plants, it is certain that enthusiasts will soon reach double figures. Where winters are wet, some early flowering species may require a little protection if the flower buds, often formed in late autumn, are not to damp off. The rewards are well worth the effort involved.

P. *alpicola*, still seen offered as P. *microdonta*, is June flowering and is one encompassing a number of varieties. The most likely species with which it can be confused is P. *sikkimensis* to which section it belongs, but a glance at the leaves will quickly provide the answer. In P. *alpicola* the base of the leaf blade ends abruptly and there is a distinct stalk, whereas in P. *sikkimensis* the blade gradually tapers to where it joins the crown of the plant. The drooping bell-shaped flowers are in umbels, each inflorescence being many flowered and carried on a scape up to eighteen inches in height. Varietal names like *alba*, *luna* and *violacea* ably describe the flower colours, but they do not indicate that the grey bloom on the leaves of some forms of P. *alpicola* var. *alba* is just as decorative as the flowers.

P. *anisodora* smells of aniseed when crushed. The flowers are borne in whorls on twelve-inch-high stems and are dark reddish-purple. This is relieved by the yellow zone that encircles the corolla. The flowers are not large and are so well spaced that no vivid display can be expected of them.

P. *apoclita* is only occasionally encountered, much less now than previously. This short-lived species must be raised continuously from seed if it is to remain in cultivation and this, of course, will depend on the enthusiasts. Its section, Muscarioides, includes many that are dainty, but in the bustle which modern life produces we seem apt to pass by these small treasures. Barely six inches high, and flowering in late spring, the deep tubular flowers, which are tightly clustered round the tip of the scape overlapping one another, have a farina-dusted ring in the throat of the tube. It is one of the high alpine primulas from Tibet and Yunnan and is completely herbaceous.

P. *aurantiaca* is one of the smaller members of the Candelabra section. It was discovered in Yunnan in 1923, and when first introduced must have created a stir.

The short scapes producing a number of whorls of orange flowers are them-selves chocolate coloured, while the leaf blade is often brightened with a red main vein.

P. aureata is the first of the Petiolarids to be mentioned. It is also one of the most fascinating and part of the story of its original introduction into our gardens is worth briefly repeating here.

In 1935 seeds were sent to the Royal Botanic Garden, Edinburgh, from the Lloyd Botanic Garden, Darjeeling, and among the packets was one of Swertia. When those seeds germinated it was observed that a 'rogue' plant was amongst them. It was a Primula, and when it flowered it turned out to be a new species, *P. aureata*. From that single plant most of the stocks in cultivation arose. It was not until 1952 that it was actually known in the wild. The *Botanical Magazine*, plate 488 (n.s.), does not do justice to the clone in cultivation. The foliage with the red mid-rib is accurate, but the petals should curve back more at the tip and the yellow centre be more pronounced. Like all Petiolarids, *P. aureata* forms its flower buds in late autumn. It is slower to come into flower in spring than some others, but even so, unless a pane of glass protects the crowns from our winter's rains, the centres are very likely to rot off. It is a perfect plant for the alpine house, but it can be just as exciting out of doors in the peat garden provided it receives proper attention. (Pl. XIId.)

P. beesiana, named after the firm of Bees in Liverpool, is a vigorous Candelabra with rosy-purple flowers. A yellow eye contrasts with the dark ground-colour of the petals. These candelabras are gross feeders and should have plenty of moisture. They are ideal for drifting among the taller rhododendrons. They can be divided in August or September, or young plants from spring-sown seeds may be planted out at the same time. Plenty of seed is usually formed.

P. bellidifolia is seen only occasionally. It is another member of the Muscarioides section, but instead of forming a spear-like spike as in *P. apoclita* the flowers radiate like the spokes of a wheel from the top of the six- to ten-inch-long scape. A green crown like a wheel cap sits on top, and the edge of this cap and the sepals are picked out with white farina. The long tubular flowers are violet-blue. To be fully appreciated it must have a place near the front. It is well named as the leaves do closely resemble those of the daisy.

P. boothii is another Petiolarid. This species has dark foliage with red backs to the leaves. As with all its allied species it tends to sit proud of the soil after a couple of years, and unless it is top dressed it will dry out. One way to safeguard against this is to lift the plants in August and September (though it is dangerous to do this if the weather is dry) and replant so that the bases of the leaves are just covered by soil. It is from these that the new roots will arise. The plants divide easily. A dome of flowers forms in the centre of the rosette of leaves and it scarcely has a flowering stem.

P. bracteosa gets its name from the unusual phenomenon of producing a young plant among the bracts which surround the flowers. It belongs to the Petiolaris section and can be induced to grow on its side in blocks of peat. In that position there is less likelihood of moisture lying in the crown and causing damage. Many of the species belonging to this section were not introduced until the 1930s. Here the large flowers are a delicate shade of lilac and again this is relieved by a yellow eye.

P. bulleyana is named in honour of Mr J. K. Bulley, the owner of the firm of Bees, who did much to finance plant collecting in China at the beginning of this century. This Candelabra has orange flowers almost an inch across, and the tiers rise at intervals one upon the other until approximately a height of two feet is reached. There is quite a bit of meal evident on the stems and leaves.

P. burmanica belongs to the same group as the last. It has reddish-purple flowers with a yellow eye, but is easily confused with *P. beesiana* which is reputed to be much more farinose, particularly on the insides of the calyces. *P. burmanica* has coarse cos-lettuce-like pale green leaves.

P. calderiana is a Petiolarid with a difference, for it has a long flower scape. It is fairly widespread in western China, Bhutan, Tibet and Nepal where it has been collected by a large number of plant-hunting explorers. This plant dies back to a bud in winter and it is only when spring arrives that the glossy leaves appear through the soil, quickly followed by the flowering scape. The large flowers are a rich shade of maroon or royal purple and again are adorned with a yellow eye. The semi-shaded part of the peat garden, or the north side of dwarf shrubs or even the peat walls themselves, are ideal sites.

P. capitata has a flowering season which lasts throughout the summer. It is not a good perennial; old plants appear to rot off at soil level, but since prodigious quantities of seed are set and germination is easy, there should be no fear of losing it. The specific name tells us that the flower head is dense and the violet flowers develop from a tight mass of buds. Although not spectacular, this species is interesting if planted in a dense group. The leaves are distinctive, being oblong with serrated margins and covered with white farina. They form a star-like rosette at soil level.

P. chionantha has pure white scented flowers. The lower scape is densely mealy and usually carries two clusters of flowers, one at the top of the scape and the other two-thirds of the way up. The Nivalis section is one of the more 'difficult' ones, but this species has remained in cultivation. The leaves stay green for part of the winter, but finally die back. In spring, when the young foliage appears anew, this is strong and upright and completely covered with farina on the undersides, a most attractive feature of this species. It usually dies after flowering, but sets generous quantities of good seed. It is one of George Forrest's discoveries. (Pl. 37.)

P. chumbiensis, from the Chumbi Valley in west Assam, is a dwarf Sikkimensis type with rich cream-coloured flowers. These are slightly farinose and are carried on a short scape, this being about eight inches tall. The leaves too are not large, but their purplish-green colouring is unique.

P. chungensis, although not in the first line of horticulturally valuable plants, contributes nevertheless to the Primula enthusiast's interest. It is smaller in growth than most Candelabras and has thin, bright green leaves on the undersides of which the veins are prominent. The open petals are orange, but a bicolor effect is given as the tube is almost red. This protrudes well beyond the small green calyx. Primula seed is known not to remain viable for long, a statement borne out in this species. Seed collected and kept over until spring usually germinates poorly, yet seeds which have fallen to the ground round the plants generally show a high breard (germination). *P. chungensis* will grow and flower quite well below taller growing rhododendrons. (Pl. 38.)

P. clarkei, although first collected in 1872, has never been plentiful in gardens. Certainly one would not expect to see large drifts of this miniature as dry seasons can deplete, if not completely kill, stocks. This tiny plant belongs to the Farinosae section of Primula. Although never more than two inches high, it forms bright little mounds of pink. The large flowers, appearing in April and May on single or multiflowered scapes, have deeply notched petals with a bright yellow eye. The small spade-like leaves are not unlike those of *P. juliae*. Division of the clumps in spring or the sowing of seed, if it can be obtained, are two ways of raising plants.

P. cockburniana is a beautiful little plant with bright orange flowers. It is a Candelabra Primula and although described as perennial is more or less monocarpic in gardens. It sows itself fairly freely, but controlled annual sowings are recommended. This Szechuan species is not known to be widespread in nature. It rarely exceeds twelve inches, has a very light inflorescence and should be given an open moist site between dwarf rhododendrons. (Pl. XIIb.)

P. denticulata appeals to a great number of people and there is no questioning either its hardiness or its colourfulness. Perhaps its greatest drawback lies in its stiff drumstick inflorescence, which gives it an artificial appearance. Because of this it is difficult to find complementary species among other moisture and peat loving plants. The choice must lie with the gardener and many colours are available. Named forms are offered, but from a batch of seedlings it is possible to select one's own strain or clone.

P. eburnea belongs to a section devoted almost entirely to rarities. Many are difficult, dying back to a resting bud in winter and producing hairy leaves in spring which can radily damp off. The section is Soldanelloides and this is one of those primulas that must be 'managed', and while alpine house treatment may ensure success, a few seedlings planted in an open, drier site near the front of the peat garden can uplift the quality. The dormant buds will require the protection of a pane of glass in winter. The flowers are pure white, almost half an inch across and are crowded into a tight head supported on a four- to six-inch-high scape.

P. edgeworthii, for years grown as *P. winteri*, is one of the easier Petiolarids. Regular division in autumn and the occasional raising of fresh stocks from seed keep this species in good health. There is no better example than *P. edgeworthii* of the forming of flower buds in autumn. Dozens of these buds crowd the crown and the whole is densely covered with meal. Discovered in the western Himalaya during the last century, it was not until around 1910 that it flowered in this country and was first described under its synonym.

P. florindae has a flowering season that lasts all summer and autumn, seed heads and flowering scapes standing side by side. It was discovered in Tibet by Frank Kingdon Ward and named after his first wife. This is a strong-growing Primula, ideally suited to the moist spongy soil among rhododendrons, and it has masses of handsome, large, spade-like foliage which completely covers the ground. Above this tower the three-foot-high inflorescences, carrying umbels of up to fifty heavily scented pale yellow flowers. There is also a copper-flowered form which has occurred in gardens and is probably of hybrid origin. This true perennial is, of course, herbaceous, but plants twenty years old can be as vigorous as those newly raised.

P. frondosa tends to be lifted by frost and so requires constant replanting during cold winters, but this upheaval in no way impairs flowering. It quickly reestablishes in spring and from the farinose buds the three- to four-inch-long obovate leaves develop. The four-inch-high meal-covered scape carries up to twenty-five rosy-lilac flowers. As these are small, one of the lower terraces would be a suitable place in which to grow this species. Its distribution is south-eastern Europe.

P. geraniifolia belongs to the section Cortusoides, which means that it resembles the Cortusa, the first member of Primulaceae to be described in this book. The foliage is Geranium-like, with irregularly notched edges and a three-inch-long hairy petiole. Long thin pubescent scapes rise to approximately nine inches from the clump of vegetation, carrying many pale purple flowers in a loose head. This species grows wild in the Chumbi Valley.

P. gracilipes is another Petiolarid which has proved easy in cultivation. It quickly forms large clumps, and it is only if these are neglected that the species fails to respond. In April the numerous large pink flowers held tightly within the rosette of leaves could be likened to a Victorian posy.

P. grandis must get its name either from the size of the leaves or the length of its flower scape, for certainly it does not describe the size of its flowers. It is a plant to be seen in a botanic garden or perhaps in the personal collection of the connoisseur and it is mentioned here only lest the unwary be attracted by its name. It is native to the Caucasus and has pale yellow corollas which are barely a quarter of an inch across.

P. griffithii is a violet-coloured Petiolarid with a flower scape eight inches tall, although this may be twice that length by the time the fruiting stage is reached. It is extremely rare in cultivation, although hybrids between it and other species in the same section are more often seen. At Keillour Castle in Perthshire, the Western home of so many of the Himalayan species, this plant flourishes. So do most of the others, and it may be an indication of the growing conditions

73

if it is recorded that the annual rainfall in the area is around forty inches and the situation where most of them grow is in a narrow gully which is kept cool and humid by a small burn. The soil is acid and very fertile, but none of this detracts from the skill of the cultivator.

P. helodoxa bears an epithet which means 'Glory of the Marsh'. Needless to say this Candelabra is well established in gardens and makes a picture when seen against a background of taller shrubs. It has strong-growing meal-covered stems three feet high producing up to six whorls of large golden yellow flowers. The spatulate leaves tend to remain erect while the plant is in its growing stage, but form a rosette once the flowering spike appears. (Pl. 39.)

P. heucherifolia has leaves which are coarser than those found in *P. geraniifolia*, but in many other ways they are similar. It also has hairy scapes, but they are only about half the size when the flowers are fully developed. These are dark purple and have rings with lighter shades near the eye. A moist shady site is most suitable.

P. involucrata is one of those plants which require very little room. The one- or two-inch-long leaves remain in a dense clump and above this is carried the twelve-inch-high stem of flowers. These are white and delicately scented, and can number up to six. It flowers in May and is best increased by dividing the clumps into small pieces in late March. (Pl. 40.)

P. ioessa is a lovely little member of the Sikkimensis group, and is most graceful in the way its lilac, funnel-shaped flowers are suspended on thin pedicels. They are also delightfully scented. There are many forms of this species available and not a few of them appear to be hybrids, but the true species needs no hybrid blood in order to increase its appeal.

P. japonica is, of course, the popular reddish-purple Candelabra with broad, pale green leaves. Although stout and carrying a great many large flowers in close whorls, the scape tends to be shorter than most in that section; it rarely measures more than 18 inches. It is so well known that little more requires to be written about it except perhaps to remind growers of two distinct forms. They are 'Miller's Crimson' and 'Postford White' and both come true from seed.

P. juliae is one of the parents of that race of primulas grown under the collective name of *P. × pruhoniciana*. To it belong the popular 'Wanda', 'Iris Main-waring' and 'Theodora'. *P. juliae*, however, is the dainty partner as it is of tiny dimensions. The leaves are more or less round, about an inch across and with toothed margins. The base of the leaf is cordate (quite different from that of *P.* 'Wanda' and its allies) and is carried on a three-inch-long stalk. The rich purple flowers are produced singly on scapes which are just long enough to carry the blooms clear of the foliage. The deeply notched petals have a zone of red close to the yellow eye, which seems to accentuate it. March and April are the flowering months, after which the plants are best divided.

P. kingii is a difficult plant; at least it is difficult in almost all gardens in which it has been tried. The only place where it appears to be at home outwith its natural habitat is in the garden of Mrs Sherriff at Ascreavie, Angus. There, treated as a bog plant, it has been a feature for years. It is a Himalayan species with dark claret coloured drooping flowers. These are carried in clusters on a short scape. In a moist part of the peat garden it is just possible it may grow. Unfortunately one must first find a source of supply.

P. latisecta belongs to the Cortusoides section, but, as the specific name suggests, the leaves are very much cut and indented. It spreads slowly like a herbaceous perennial and can be treated in the same way when dividing becomes necessary. The leaf stalks are bristly with hairs and are often red, while the wiry stems, rising six inches above ground level, carry the rosy-pink flowers in few-flowered umbels. It is happy only in a shaded part of the garden.

P. luteola is from the Caucasus. Its yellow flowers appear in summer and are carried in a close umbel of twenty flowers or more on top of eight-inch-high stems. It certainly enjoys moist conditions in summer, but is unhappy if kept too wet in winter. The dentate leaves, lanceolate in shape, can be as long as the scape is high.

P. macrophylla is a Nivalid. That means it is not one of the easiest to grow and in cultivation usually dies after flowering. It is one of the smaller species which may carry twenty flowers in a head, but usually far fewer, the stems sometimes being only six inches high. The plant is well dusted with farina, particularly the undersides of the leaves, which are completely covered. The purple flowers, measuring almost an inch across, have yellowish eyes.

P. megaseaefolia is rarely seen despite the fact that it is most attractive. The specific name is well chosen, for the foliage is very like that of Bergenia, or Megasea, as it was at one time known. It is an evergreen which no doubt accounts for its being recommended as a cool greenhouse plant, but in the shelter of a Rhododendron, shaded from the sun in the same way as one would treat *Epigaea gaultherioides*, it can be grown out of doors and produces its pink flowers on five-inch scapes in early spring. It was discovered and introduced from Lazistan at the beginning of this century by a very famous plantswoman, Miss Ellen Willmott.

P. melanops very closely resembles *P. chionantha*, except that here the colour of the flowers is deep violet to rich purple. In addition to this a zone of black forms the eye. It blooms during May and June and sets ample seed for future generations. The leaves are lanceolate, although the recurved margins give the impression that the foliage is much narrower than it really is, while the undersides are densely covered by cream farina.

P. muscarioides could not have a more descriptive specific name. George Forrest said of this plant that it inhabits moist, open, grassy situations on the margins of pine forests. Surely this immediately suggests a suitable inhabitant for the peat garden. Its tight inflorescence of pendent blue flowers is capped by the circle of deep purple formed by the calyxes. The light green leaves are up to five inches high and for most of their length taper towards the base. It grows wild in Yunnan, but flourishes here provided seeds are saved and sown annually.

75

P. nutans, Soldanelloides section, is a gem among Primulas. It was first discovered by the Abbé Delavay, who christened it 'nutans' because of its nodding flowers; but it was not until 1916, when George Forrest sent seed home, that it was seen in Britain. The flowers are most attractive, being delightfully scented and pale violet in colour and carried on twelve-inch-high stems. Inside, the corolla is dusted with white farina, as is the stem. Even the foliage, which is fairly erect, has curled-back margins which protect the farinose undersides. It grows easily in gardens and, although a percentage of the plants may show perennial tendencies, it is mostly monocarpic in cultivation. As abundant seeds are formed and these germinate freely there is little excuse for losing it. (Pl. 41.)

P. obconica hardly seems a plant to mention here, but there is a form known as var. *werringtonensis* which is hardy or nearly so. Forrest records this as a small form of *P. obconica* which was discovered in Yunnan. The flowers are reddish-purple, the leaves slightly crenate and the name *werringtonensis* denotes a garden in Cornwall.

P. obtusifolia, a Nivalid from the north-west Himalaya, with purple bell-shaped flowers measuring at least an inch across, is well worth searching for. Up to three whorls of flowers may appear on a stalk, but it is usually the terminal one only that develops. The stems are generally twelve inches in height while the leaves, rounded at the point (the character which gives this plant its second name), are approximately six inches long. Annual seed sowing is necessary to keep this species in the collection and, like all members of this section, the level of planting should be the same as it was in the frames carrying the pricked out seedlings. If too deeply planted the stems are liable to damp off at the neck.

P. polyneura is one of the most useful plants for growing in shade. There it produces a complete soil covering with its thick, downy, scallop-edged leaves; in fact the undersides are white with netted pubescence. Above this the flower scapes display their heads of large flowers, and these may vary from pale rose to deepest purple, all with a yellow eye. (Pl. XIIa.) A form listed as 'Highdown' bears almost magenta flowers and in May as well as throughout the summer certainly brightens up a darkish area. They are long-lived perennials and may be increased by division in early spring. It is the finest of the Cortusoides section and was discovered by Ernest H. Wilson in west Szechuan around the beginning of this century.

P. prolifera provides a wonderful display in May and June. It is then that its tiers of golden yellow flowers are fully open and, standing two feet high, the scapes are proud and erect. This Candelabra is without farina and has oblong crenate leaves of twelve inches or more which taper towards the base. It is native to Assam, but it is completely acclimatized to the conditions here.

P. pulverulenta is one of the most popular Candelabra primulas, and deserves to be because of the influence it has had in bringing colour to our gardens. It is a plant much covered by meal, in fact the two-feet-high scape is pure white, a factor which made it easy to name. It has whorls of large, deep crimson flowers each with a purple eye, and grows equally well in shade or full sun, the main criterion being that there is sufficient moisture. It was introduced around 1905 and since that time it has freely hybridized with a number of others belonging to the same section. Its influence is always apparent when the amount of mealiness on the progeny is observed, but the myriad colour forms that now exist in unselected strains, many of which have been named, could be likened to a rainbow.

P. reidii is another rich treasure. Soldanelloides is its section and it requires the care suggested for *P. eburnea*. No one could say that plants belonging to this section are easy in cultivation, in fact they are very difficult and, while some may be virtually impossible to establish in some gardens, it is only by enthusiasts experimenting with them that progress towards their better understanding will be made. Moisture on the dormant buds spells anathema to them. *P. reidii* is a most beautiful little plant with scapes reaching four to six inches topped by a cluster of pure white, wide-open globular flowers. These flower heads may carry very few blooms or as many as ten, while at ground level the hairy leaves form a rosette at their base. This species is not long lived, but it can be kept in cultivation by successive seed sowing.

In recent years an apparently more vigorous variety has been introduced. It has blue flowers, although much of the interior of the bell may be shaded white. The flowers change abruptly from a narrow tube to a wide-open corolla, the petals of which are incised. But what is even more delightful is the scent, of which one is much more conscious, of course, in a confined frame or alpine house. This variety is known as *P. reidii* var. *williamsii* and was collected in Nepal during the plant-hunting expedition by Stainton, Sykes and Williams in 1954. (Pl. XIIe.)

P. reticulata, a small member of the Sikkimensis group, demands a site near the front of the peat garden. Rarely do the scapes exceed nine inches and they bear a number of pendulous, creamy-yellow flowers. It is a graceful species with leaves showing deeply marked venation. The foliage may be reddish on the underside with leaf blades up to three inches in length, but it is a plant which is neither vigorous nor invasive. Collected on the Nepalese-Tibetan border, it has been in cultivation for almost a century.

P. rosea, that bright, early flowering, rich rose-coloured species from the northwest Himalaya, is well known to gardeners and provides a vivid display if planted in a moist site. The soil too should be rich, for only on a well-tilled fertile loam will most of these gross feeders prosper. It dies down completely in winter, but from the moment the buds begin to expand its presence is apparent. The very small leaves are coppery-red while the flower buds, which start to open right in the heart of the plant, add brightness from the start. Eventually the scapes are four to five inches in height. Forms with particularly vivid and large flowers, offered under clonal names, are worth cultivating. (Pl. 42.)

P. scapigera was discovered by Hooker in 1848 in Sikkim. It belongs to the section Petiolaris, about which so much has been written yet never enough. This group will always both attract and frustrate the gardener. Our species is one of those easier to grow. It is a true perennial and proliferates freely. At flowering time, which is March and April, a dome of large pink flowers completely fills the crown and the leaves take the form of a collar or ruff. The petals have a frilled edge, while the greenish-yellow centre is enclosed in a white circle. (Pl. 43.) *Primula* 'Pandora' is one of its selected hybrids with *P. edgeworthii*, although the group name *P.* × *scapeosa* refers to the progeny when crossed with *P. bracteosa.*

P. secundiflora is unique among primulas in that the flowers are supposed to be arranged on one side of the stem. This is more or less true. It is a distinct Sikkimensis type of flower, up to three-quarters of an inch across and reddish-purple in colour with a slight bloom. The stems may carry as many as twenty flowers and are usually little more than a foot in height, while the serrated leaves, which can be as long, tend to lie flat at flowering time. This is a most desirable plant from Yunnan and Szechuan.

P. serratifolia, a Candelabra with a difference, is perhaps less vigorous than many, but has most distinctive flowers. These are arranged in tiers in the same way as all the others, but the individual blooms are yellow with an orange star decorating the centre. Plentiful on the Tibetan-Burmese border, it is a great pity the same cannot be said of it in cultivation. (Pl. 45.)

P. sieboldii, much used for conservatory decoration, has its place in the open garden. Very wet sites may cause it to damp off, but it is happy in many gardens. The leaves develop quickly and the blades remain stiffly vertical as they elongate. It is only later when flowering takes place that they assume a more normal plane. May and June are the main flowering months, and it is then that the full benefit can be got from the very neat, large, round, white-eyed blooms which are carried in up to ten-flowered clusters. It tends to spread, almost sucker, so that established clumps can be extremely floriferous. Many horticultural forms have been marketed and these vary from pure white to rose, purple and even bicolor. (Pl. XIIf.)

P. sikkimensis was written about in the *Botanical Magazine* of 1851 and its beauty was much appreciated then. Horticultural writers still extol its virtues, among which are that it is a good perennial, has bright yellow flowers which are nicely scented and it remains in bloom for a long time. This is the species which gives its name to the section. It has a fairly wide range, being wild in Nepal, Burma, Sikkim, Yunnan, Tibet and Bhutan, which more or less tells the grower that there may be good and better than good forms. A packet of seeds usually supplies this range. An especially large-flowered form was introduced by Oleg Polunin in 1949 under his number P 97, but this may not still be in cultivation.

P. sinopurpurea, a Nivalid, is a most striking plant. It deserves a prominent position where its leaf arrangements and flowering stems can be admired. The fairly broad toothed leaves have an almost suède texture while below they are covered with yellow farina. The scape is likewise dusted, but on top of this, and perhaps three-quarters of the way up, clusters of pinkish-purple flowers are carried. These flowers have a prominent white eye. Although perennial, they tend to die after flowering, but they produce a generous seed harvest. George Forrest first collected this species in Yunnan almost sixty years ago. (Pl. 44.)

P. smithiana has bright yellow flowers, small when compared with *P. prolifera* and *P. helodoxa*, but none the less attractive. Its leaves too are not so coarse, the scape is finer and, overall, it is perhaps more subdued. Like other Candelabras it flowers in June.

P. sonchifolia is an aristocratic plant which fortunately is well established in some northern gardens. The peat-wall environment is the correct one, for it is only in a moisture-laden atmosphere that these plants will survive. Native to the high mountains of Yunnan, in fact referred to by Forrest as a high alpine, and one of the first to show as the snowfields diminish, it overwinters in the bud stage; that is to say all that remains visible in winter is a tight bud of leaves in the core of which the undeveloped flowers rest. As spring progresses so the leaves open and expand and, contrary to many other Petiolarids, a distinct scape carries the flower head approximately three inches clear of this foliage. The bluish-purple flowers showing a white ring with yellow centre appear in March and April.

P. strumosa is yellow flowered. It could be likened to a yellow form of *P. sonchifolia*; it belongs to the same section but is more easily grown. It is also later in flowering, May and June being its season to bloom.

P. tsariensis, without doubt of great garden value, but, alas, also a Petiolarid, has rich bluish-purple flowers with a most pronounced yellow eye. The floral umbel may consist of four or five flowers and be carried on a six-inch scape. The smooth tapering leaves are dark green. Semi-shade and the avoidance of a dry atmosphere will help towards its successful cultivation. (Pl. 46.)

P. viali is unique, for the scarlet unopen buds are reminiscent of the red-hot poker. The open flowers, however, are pale violet and are carried in a tight spike-like arrangement. The inflorescence can be up to six inches long on stems eighteen inches high, but this is exceptional. The lanceolate leaves are extremely downy and are almost as upright as the flower spike; invariably the measurements are less. Although perennial in nature, most plants die after flowering in cultivation, and there is no doubt whatever that the annual raising of young plants, which will flower the following year (i.e. treating this species as a biennial), is the proper method of dealing with it.

P. vulgaris is our own native primrose and it would be a valuable addition to any garden but, like so many desirable British native plants, it often shuns the company of aliens. *P. vulgaris* subsp. *sibthorpii*, however, a form from the Caucasus, can be induced to remain, and the pink flowers, which develop before those of the yellow native one, contribute to the early display.

P. waltonii, a Sikkimensis type with wine-coloured flowers, opens its buds in May and June. There is a range of shades involved, not all of which are attractive, and, because of the free manner in which the pollen is dispelled, hybrids which are most undesirable often appear after fresh plants have been raised from seed. Gardenwise it is not of high merit.

P. whitei perpetuates the name of Sir Claude White, Political Officer in Sikkim, who collected seeds of this species, new at that time (1905) in Bhutan. More recently a plant in many respects similar, but at the time considered to be botanically different and certainly horticulturally superior, was collected by Ludlow and Sherriff and named P. bhutanica. Later research proved that they were one and the same species, and as the original name was P. whitei it took precedence. No doubt Sir Claude deserves the honour, but what a pity so romantic a name has to be submerged. The praises of this, one of the most beautiful of flowering plants, cannot be oversung. The ice-blue flowers tightly packed within the rosette of coarsely toothed leaves command admiration. No one who sees the flowering plant in March can walk casually by. It rewards the cultivator who provides the pane of glass which gives the winter protection required in some areas, though this may not be necessary where weather conditions are more severe and temperatures at that time do not fluctuate. In winter the plants are reduced to buds, for all the world like Brussels sprouts, but with a different end product! Among all Petiolari primulas, and these are in the region of fifty, P. whitei (bhutanica) ranks as one of the finest. (Pl. XIIc.)

P. yargongensis is the pinkish-lilac counterpart of P. involucrata. In fact it was at one time known as P. involucrata var. wardii as well as P. wardii. There seem to have been much confusion and uncertainty about its naming, but at the present moment the colour appears to be the deciding factor. From a garden point of view they certainly are different and both are worth growing. This pink species looks well planted in a drift and, what is more, it is pleasantly scented.

TRIENTALIS is a genus of two north temperate species and is highly prized in both Europe and North America where it grows wild. Both species grow in pine woods among the other forest-floor plants and there is no doubt that the underground spreading parts revel in the looseness of the mossy, Vaccinium-type of association. Plant neighbourliness plays a large part in their lives.

T. borealis, the North American plant, is better described as "Star-Flower". It is more robust than its European ally and has coarser leaves, but it does not have so distinguished a flower. It blooms in summer at the same time as the British native species.

T. europaea is the one on the east side of the Atlantic which is widespread in northern Europe. The thin wiry stems stand above the undercover heath, but in the peat garden it will still run through the roots of rhododendrons. From a whorl of lanceolate leaves, the white, starry flowers, usually borne singly on fine threadlike stems, seem from quite close at hand to be unsupported. An inappropriate common name sometimes used is "Chickweed Wintergreen".

9 THE LILY FAMILY (*Liliaceae*)

BULBINELLA has tuberous roots, or, to be more precise, the herbaceous stems die back to thickened radiating roots which spread horizontally as well as deeply into the soil. There are a few species spread over South Africa and New Zealand, but only those native to New Zealand are hardy enough to be cultivated out of doors in our gardens.

B. *hookeri* is one of the easiest plants to grow; in fact self-sown seedlings can become a nuisance. The bright yellow flowers, appearing in early summer, are carried in densely flowered racemes. These flowering stems may reach two and a half feet in height. In early spring, as the foliage emerges from the soil, it has a distinctly bronze appearance, much of which it retains.

B. *rossii* is a good garden plant. It was discovered near Auckland, New Zealand, around the middle of the last century, and although it is said to grow in drier sites than *B. hookeri* it does not appear to object to growing in the peat garden. Unfortunately it is not nearly so easily obtained as the former species but it is the more desirable. Its flowers are a bright shade of yellow and are larger than *B. hookeri*; the flower spike, although not so elongated, is much broader.

CHAMAELIRIUM is closely related to Helonias and was once known as Veratrum, but it is such a distinct plant that it has been placed in a genus of its own. It is not of great garden value, but in a collection of plants it has its place.

C. *luteum* is the sole species. It is herbaceous. dying back to a crown, but producing in spring a cluster of small Veratrum-like leaves. It likes a soil rich in humus and in most years will produce a spike of white flowers which fade to cream. The individual blooms are tiny, but so tightly packed are they on slender spikes up to nine inches high that they look more like tails! The flowering period is long, lasting from May to July.

CHIONOGRAPHIS is native to Japan and, according to recent Japanese floras, includes a number of species, but only one is ever seen in cultivation. It belongs to a most interesting group of low-growing herbaceous perennials which in nature occur in moist areas, sometimes by streams or in cool, open woodland. Chionographis dies back in autumn to a stout rhizomatous rootstock.

C. *japonica* is the species grown in this country. Its bright green leaves form a rosette at ground level while the twelve-inch-high flowering shoots carry on the upper part a collection of evenly spaced flowers. The colour of the thin, strap-shaped petals is dazzlingly white, while on the lower part of the stem and spaced at wide intervals a few leaves are borne. Seed sowing or dividing the dormant crowns are ways to propagate this plant.

CLINTONIA is a genus containing only a few species. It is an interesting one, not because of its flower power, which is neither garish nor plentiful, but for its delicate floral arrangement and the fruits that follow. It is completely herbaceous in growth, dying back in at least one species to a dense cluster of crowns an inch or so beneath the surface of the soil. Others tend to spread rapidly, becoming a nuisance in fact, but fail to flower freely. They certainly enjoy the loose puffy soil of the peat garden. Once established those that build up slowly are best left in peace since not all take kindly to transplanting. Raising young plants from seed is the best way to increase stocks, although in my experience this genus does not seem to produce very many seeds in gardens.

C. *alpina* is a small plant six inches or slightly more in height. It has broad semi-folded leaves which are dull green on the upper surfaces and tinged with red on the undersides. The scape is terminated with a cluster of drooping white flowers, but with a bluish sheen to them. Its natural distribution is wide, spreading from the Himalaya into eastern Asia.

C. *andrewsiana* is surely the finest member. It is a stately plant, much admired by plantsmen. In summer, from a compact cluster of crowns, it produces a number of broad, bright green leaves, not unlike those associated with Cypripedium. From the centre of these arises the single flower stalk, in some seasons growing to two feet. A number of pinkish-purple, bell-shaped flowers terminate this stem with one or two subsidiary clusters occurring lower down. Although every crown does not send up a shoot, well-established plants can be relied upon to produce a few. In autumn bluish-purple berries appear, but few good seeds seem to form.

C. *borealis* is an attractive miniature with broad ciliate leaves reminiscent of those of the "Twayblade". The thin stems carry few-flowered umbels of greenish-yellow bells with reflexed tips. It is a most attractive species from the eastern states of North America.

C. udensis is treated as a species by the Japanese botanist Ohwi, although others considered it a mere form of *C. alpina*. It takes its name from one of its wild stations in Mongolia, though it also occurs on a number of Japanese islands. Flowering in early summer, and native to coniferous woods, it enjoys the same conditions in cultivation. The lax head bearing few flowers is carried on a twelve-inch stem. The white flowers are followed by nearly black fruits, but here also few seeds are produced in gardens. Colonization can occur by means of creeping rhizomes.

C. umbellata is well described and the fairly dense umbel of white flowers is quite pronounced. The broad leaves are typical of the genus, three or four clothing the tip of each rhizome. This is a North American species which increases fairly satisfactorily when conditions suit it. The flowers are more than just handsome, they are also scented; berries are not uncommon later in the year.

C. uniflora may suggest to the reader a sparsely flowering plant, as indeed it is, but it makes up for the shortage of blooms by producing flowers which are almost white, probably as large as those of any other species. They are usually produced singly, although, occasionally, two may be seen on the same stem. The plant occurs in coniferous mountain forests and is affectionately known in California as "Bride's Bonnet". Of all the species this one spreads more than the others do in the sandy soil at Edinburgh and perhaps, rather than include it among the rarities, its site ought to be confined to the neighbourhood of less valuable plants.

DANAË is a shrubby plant with evergreen foliage which is much used for background and contrast in floral arrangements. It is particularly plentiful in winter when this type of material is in demand. This genus, containing one species only, is akin to Ruscus, "Butcher's Broom". It is a plant for woodland, semi-shaded conditions where a colony formed by its suckering shoots will soon be made. Its shoots are virtually biennial, dying completely, so that it is necessary to remove the dead stems regularly from the clumps to keep the plants tidy.

D. racemosa is the "Alexandrian Laurel" and its broad leaf-like structures are in fact flattened stems, coloured green like leaves, and functioning as foliage. The infinitesimal leaves can be seen in the centre of some of these broad stems and in their axils the flowers occur. Although in the eastern Mediterranean countries, where the plant is native, the red fruits must be colourful and attractive, it is sad to relate that very few berries are ever seen in this country.

DISPORUM is a reliably hardy herbaceous perennial. In some ways it is similar to Uvularia. It is native to Asia, including Japan and a number of Pacific islands, and western North America, and although there are said to be more than fifteen species not many are seen in gardens. Those listed here are interesting for both their foliage and fruits and may be planted not too far in the background.

D. pullum is an oriental species stiffly upright in growth, the upper parts of the stem arching over and being well furnished with foliage. Clusters of white to purplish tubular flowers form in the axils of the upper dark green leaves and, if fertilization takes place, some of these may be followed by red berries.

D. smithii is a Californian native with yellow, string-like, deep searching roots, making these plants difficult to dislodge. This is particularly stressed here as a warning lest seedlings—and they can be many—germinate in the crowns of rare plants; such seedlings should be removed at an early stage. From over-wintering crowns the wiry stems ascend, bearing attractive bright green foliage. The upper leaves protect the flower clusters, and these are invariably followed by quite large orange-yellow three-sided fruits. These fruits are of long duration, making this species prominent and decorative for much of the autumn. (Pls. IId and 10.)

D. trachycarpum, also of western North American origin, while less attractive than the previous plant, is worth including where a collection of liliaceous species is grown. It flowers continuously throughout the summer, and this plant can be said to display its creamy-white blooms as well as any of the other species. It will reach two feet in height and although it does not produce many fruits its long season of bloom in some ways compensates for this.

ERYTHRONIUM includes a number of beautiful plants which, apart from the sole European member (E. dens-canis), are of North American origin. They possess a number of attractive characters which include marbled patterns on their leaves, giving rise to the common name "Trout Lilies". The elegant way in which the flowers are supported as well as their not too tall or small stature adds to their popularity. "Avalanche Lily" is another name by which some white forms are colloquially known and, while it is stated that the reason for this name is probably the way these plants quickly follow the receding snows, I sometimes wonder whether it is not also because the flowers en masse resemble an avalanche. Attractive as they are, they die back in summer, so planting to excess could leave bare areas. Sparse planting through non-competitive genera like autumn-flowering Asiatic gentians (G. sino-ornata), or in the gaps between dwarf rhododendrons are suitable places. Transplanting in late summer should be done quickly as the long tuber-like bulbs ought not to remain exposed to the atmosphere, where they will dry out and go soft.

E. americanum is one of the smaller species, the flowering stems rarely exceeding four inches in height. The flowers are golden yellow with darker markings on the outsides of the much reflexed petals. Colonies tend to become crowded, but even then, although flowering may be impaired, the mottled foliage is attractive for a while.

E. californicum grows on slopes and in woods in its native California, but fortunately for us its habitats are high in the mountains, so ensuring complete hardiness in this country. Two or three cream-coloured flowers are borne on nine-inch-high stems. If the bulbs are planted approximately three inches deep they will multiply quite quickly.

E. dens-canis is our own "Dog's-tooth Violet". It occurs throughout Europe, in Asia as far east as Japan, and a few selected variations are offered. White and reddish-purple blotches mark prostrate broad leaves, on top of which the short frail flower stems support the blooms. Each stem carries a single flower and these can be white, violet, violet-purple or rose-purple, according to the variety grown. Late March sees this species in bloom.

E. hendersonii is an aristocrat from south Oregon. The stiff reddish-purple stems, up to a foot high, bear several pendent flowers. These are violet purple apart from the bases of the petals, which are dark purple. April is the flowering month of this species but the leaves are attractive too, being irregularly shaded with green and purple patches.

E. oregonum is often said to be a vigorous form of *E. californicum*. It would be difficult to say which is the better. Some tall flowering stems are recorded up to two feet, but much more modest measurements are the rule. It is this species which is associated with the cultivar 'White Beauty', a prolific hybrid with large creamy-white flowers. Unless frequently divided the grace of the individual stems is lost in the mass.

E. revolutum, one of the most beautiful members of this very fascinating genus, cannot have too much said in its favour. It is a western North American species, handsome and proud, bearing two or three large flowers on ten-inch-high stems. The drooping flowers, usually pink in colour, are held clear of the mottled foliage and the main flower stem so that their well-balanced proportions are displayed. The fact that it does not increase at the rate of some others allows it to be enjoyed longer before division is contemplated. A number of named forms have been segregated, 'Pink Beauty' and 'Johnsonii' being two. It is one of the last species to bloom, May being its flowering month. (Pl. 11.)

E. tuolumnense is a Californian species with a difference. Instead of having leaves which are heavily patterned they are light green, broad and are held more upright on their petioles. The flowers are golden-yellow, complementary to the pale foliage. Although small when judged by flowers of other species, their lightness of colour makes them stand out in an area which is partially shaded.

FRITILLARIA is a large genus and, although many species native to the eastern Mediterranean are not generally considered as peat-garden plants, it has amongst its many forms a few which seem to enjoy the company of peat lovers. In order that the range of worth-while plants should not be too diluted a few of these are considered here. The bulb enthusiast is, of course, bound to say that these are the easily grown ones, but this fact can be advantageous.

F. camschatcensis occurs in western North America, the Aleutians, Kamchatka and Japan, and although not robust or easy adds another interest. Its foliage is light green with well-defined veins, the lower leaves being arranged in whorls round the stem while the upper ones are alternate. The flowering stems are usually nine inches or so in height and carry up to three dark purple, almost black, bell-shaped corollas. The green stigma and yellow anthers brighten the centre of the flowers. Multiplying bulbs tend to reduce flowering, so frequent lifting and replanting of the larger-sized bulbs may have to be done in order to retain this species in vigour. It so closely resembles the characters that signify Lilium that some authorities place it within that genus.

F. meleagris barely requires description here, but it is listed because of its value in the garden. It is easy to grow and, because of its narrow grass-like leaves which, unlike its seed-bearing stems, return early to rest, can be grown through other ground-covering genera with different flowering times. The chequered petals, be they dark purple or white, never fail to create interest and admiration. The manner in which the flowering stems appear to come straight through the centre of the divided bulbs is something the gardener notices as he handles them. (Pl. IIf.)

F. pallidiflora is a hardy species from south Siberia. It is fairly robust, growing to a height of eighteen inches or thereby and carrying at the top of its shoots a number of wide, bell-shaped, greenish-cream flowers. They are faintly spotted within and in size completely unexpected, being at least one and a half inches long. Add to this a long season of bloom, when compared with other spring bulbs, broadly lanceolate foliage which is densely glaucous, and you have a very attractive plant.

F. pyrenaica is not the most decorative of fritillarias, although there are many others which could never be classed as attractive, but a clump of this species can provide a feature. It flowers in early April, usually two being carried on each stem. The colour of the perianth may vary from dark purple to dark green and in fact a yellow form has been recorded, but this is a species well worth procuring whatever the shade.

F. roylei is a Himalayan plant of easy cultivation. It is completely covered in glaucous meal, an addition to the chequered multi-shaded flowers, thus presenting an unhandled look. Two or three of these one-and-a-half-inch-long bell-shaped blooms constitute an inflorescence and the flowering stems are usually in excess of twelve inches.

GAGEA, of which there are in the region of forty species, has little of horticultural merit to recommend it. However, there is at least one, a British native plant, which may appeal to the bulb grower, and because of its early flowering habit brings a little colour and interest to the garden in March.

G. lutea, as the specific name implies, is a yellow-flowered bulbous perennial. Its common name, "Yellow Star of Bethlehem", indicates its flowering habit. It usually reaches a few inches in height, but quite often, particularly early in the season, the flowering stems seem scarcely above ground level before the first flowers are showing. Its trouble-free nature and harmless manner of growth give no reasons for excluding it from an interested gardener's collection.

HELONIAS is a genus of a single species which has been known as "Swamp Pink", although foliage-wise the leaves do not resemble our native "Sea Pink". It grows in moist areas in east North American states and flowers during May.

H. bullata is an evergreen plant which has an over-wintering crown of strap-shaped, greenish-yellow foliage. From the centre of this circle of leaves the twelve-inch-long flowering spike emerges. This is topped by a tightly packed head of small pink flowers with contrasting greyish-violet anthers and it is these which give the flowers their hazy aurora. If happy, the plants will produce offsets, and dividing and replanting these clumps in spring is an easy way to increase stocks.

HELONIOPSIS is not unlike Helonias in foliage, and in winter one could not be blamed for confusing the two genera. It is not so robust as Helonias, but usually a tight bud is prominent in the centre of the ring of leaves during the dormant season. There are a few species, all those available being sought after as they have that elusive Japanese quality so much in demand by gardeners. The plants are rhizomatous.

H. breviscapa is a dwarf species. It flowers in May, the flowering stem rising quickly from a tight winter bud. The stems carry a few strap-shaped leaves, but it is the relatively large pink flowers, albeit few in number and forming short racemes, that catch the eye. The inflorescence appears to be spiky, but in reality this is due to the wide-open flowers allowing the stamens to protrude far beyond the petals.

H. japonica, in some books referred to as *H. orientalis*, is quite distinct immediately it begins to flower. Whereas *H. breviscapa* has a stiff inflorescence, in *H. japonica* the flowering stem curves and it is on this arched part that the flowers occur. They are pink in colour, but the petals do not spread so widely as do those of *H. breviscapa* and so the tubes of the blooms are obvious. It grows in mountainous areas in thickets and meadows at high altitudes in many of the Japanese islands.

LILIUM is the plant which gives the family its name and, if the christening of a family had to wait until all the genera in it were discovered, it is doubtful whether any other genus would have been chosen. Lilies are among the most interesting and decorative of plants. They have a wide distribution which extends round the northern hemisphere and, with the variations of climate in the circumpolar spread, obviously the morphology of the flowers, superficially at least, is different. Wide, open flowers are found in *L. auratum*, trumpet-like blooms in *L. regale* and, in the martagon types, flowers in the form of a Turk's cap. Height, too, varies from a few inches to eleven and twelve feet. However, I am dealing here only with those which are suitable in size for including in the peat garden, and it is fortunate that this number is not too small. It includes species indigenous to America, Europe and Asia, and these embrace all the flower types.

Only species are dealt with, as man-made hybrids lose their popularity after a few years as well as deteriorate through continuous vegetative propagation.

Lilies have always been considered difficult, and while this is no doubt true in a number of cases it is wrong to get the impression that this holds for all species. Nevertheless our gardens would be very much poorer if we were to restrict our choice of plants to those which required the minimum of care, and so no attempt is made here to select the species accordingly. As always, the criteria are interest, colour and form, of which interest must be given a great deal of priority.

The conditions in a peat garden are right for most lilies, although a little extra attention in the form of additional soil preparation and plant hygiene can raise their quality and subsequently the pleasure derived from growing them. Many lilies are long lived and, what is more, if established in good soil may remain undisturbed for many years. Other species, because of their method of bulb proliferation, must be handled regularly, i.e. every few years if their standard of vigour is to be maintained. It is among the former, fortunately, that most of those in this list belong. Rapid drainage is very important.

L. amabile, a Korean species, has flowers of the Turk's cap type which are deep orange in colour. They are heavily spotted, with only a few flowers on the stem, and have broad grass-like foliage. It is usually seen standing three feet in height and like quite a few fine lilies it is the better for being supported by a thin stake. Beware when inserting these in the soil lest the cane is pushed through the bulb.

L. bolanderi, hailing from North America, is rarely seen except in the gardens of specialists. It is a tiny lily, so much so that in Woodcock and Stearn's book, *Lilies of the World*, they refer to it as the "Thimble Lily". This is a very apt description as the flower, little more than an inch in length, could well be likened to one. Up to nine flowers are recorded as being produced on three-feet-high stems, but generally the number is fewer and the height less. The blooms are brick-red in colour, spotted on the insides, while the small leaves arranged in whorls on the stems have a slightly glaucous hue.

L. canadense is rightly classed one of the most beautiful of lilies. It is unfortunate that in cultivation it is reluctant to remain for long, despite the ease with which it appears to spread in the wild. Luckily bulbs are regularly offered for sale; these are small and rhizomatous in habit and often two growing points are received on one so-called bulb. The orange flower is tubular or bell-shaped at its base, while the tip of the petal turns back but not so severely as in *L. martagon*. It has purple spots in plenty at the base of the tube. So graceful is this species that viewed from a distance these delicate blooms appear to be suspended without support, for the flower stalks are so fine. Both yellow and red forms are sometimes seen and usually carry the varietal names of *flavum* and *coccineum*. (Pl. 19.)

L. cernuum grows on the mainland of eastern Asia in Manchuria, Korea and the U.S.S.R. Its name describes the drooping habit of the flowers. These martagon-type flowers, numbering six or more, top the stiff wiry stems two to three feet high. They are pinkish-purple in shade and have deeper coloured blotches. Numerous narrow leaves surround the stems.

L. columbianum is quite widespread in its home states, but on the whole its stay in gardens is not of long duration. The regular raising of fresh stocks from seed is recommended. This is a slender, graceful lily with well-spaced bright orange flowers of the Turk's cap type which are held well clear of the whorls of leaves. These often have a wavy margin and a slight suggestion of meal. The accentuated reflexing of the petals brings the yellow stamens into prominence.

L. davidii is one of the easier lilies and also one which benefits from regular top dressing with bulky compost. This goes for many of the Asiatic species as most of them have feeding roots which emerge from that part of the stem below ground level. This supplementary root system is annual; it helps to boost supplies of food and moisture to the developing shoots and dies when they wither in late autumn. Bulbils sometimes form on that part of the stem covered by soil, and after a year these young plants are better lined out in another part of the garden to grow on. They then do not have to compete with the parent bulbs for the available sustenance. Numerous grass-like leaves adorn the shoots which are in turn terminated by a many-flowered inflorescence of reddish-orange martagon-type flowers.

L. grayi is in many ways similar to *L. canadense*. Its main difference lies in the way it holds its flowers. These are only partly pendent, in fact their plane is almost horizontal, and the flower tips are hardly reflexed at all. Stems four feet high are normal, if one can say anything is normal about a rare species, and although they may seem a trifle tall for some gardens it is as well to bear in mind the daintiness of lilies generally; their sparse foliage and overall lightness influence their neighbours virtually not at all. The flower colour is bright red. (Pl. VIa.)

L. henricii is included here for the record. It is unlikely that it will ever be available in a general way. In fact at this stage, apart from a few seedlings raised from seeds distributed from home harvested seeds, there is only one colony known in cultivation. George Forrest collected seeds in 1919, and this material stems from that harvesting of over fifty years ago. It is located in a much-visited private garden in Perthshire where lily growers pay homage whenever they can. Late June to August is the period when most lilies bloom and this one fits into this pattern. The open hood-like flowers, akin to Nomocharis in shape, are pure white apart from the deep reddish-purple blotch at the base of each petal. If *L. henricii* were easy and plentiful it would still be much admired.

L. japonicum is known in Japan as the "Bamboo Lily". This may be because it grows among dwarf bamboos in nature, but if its slender stems were compared with bamboo shoots this would afford another reason for the popular common name. This species is held in high regard by lily enthusiasts, being one of the most beautiful plants in cultivation. The lanceolate foliage is decked with grape-like bloom. Like most delicate things the flowers are small in number, but their size is quite large, being four to six inches in length. They are funnel-shaped, pale pink in colour and for contrast bear anthers with pollen which is dark orange-red. Light partial shade is necessary to retain the delicate tone of the flowers. Once established *L. japonicum* will flower for many years in the same site. (Pl. VIb.)

L. lankongense is a traveller, that is to say it spreads, and this is due to the bulbs forming on the laterally spreading underground shoots. The leaves are dark green with a purplish tinge which seems to blend in with the purple-spotted, rose-pink flowers. These are martagon types, few in number and carried on stems up to three feet in height. Although it spreads it is not troublesome in the same way as *L. duchartrei*, which can colonize an area to the exclusion of all else, even its own flowering-size bulbs.

L. mackliniae was discovered only in 1946 on the borders of Burma and Assam near Imphal, and is sometimes referred to as the "Manipur Lily". It immediately became popular and not without cause, for it is easily raised, flowers quickly from seed and remains in excellent health for a number of years. Like *L. henricii* it bridges the gap between Lilium and Nomocharis and also, like that species, has the open hood or umbrella-type flower. This is pale pink on the inside and light pinkish-purple on the outside. Usually only two to two and a half feet in height, its nodding habit prevents the delicate shade of the insides of the flowers from being fully appreciated. The pale green stigma, prominent in the centre of the flower, seems to command attention.

L. martagon could hardly be classed as a good garden plant, although a well-grown clump bearing numerous stems of dull purple flowers can provide a feature. It is widespread in Europe and occurs in Britain, being one of the first to flower. It has a spire-like inflorescence which gives a stately appearance. Among its inevitable variations are one or two well worth considering, and the cultivar *L. martagon* 'Album' is one of the finest white flowers to be found in Lilium, in this instance being devoid of spots. *L. martagon* var. *cattaniae*, on the other hand, so dark in colour to be almost black, needs a light background to set off its flowers. Multiplication by bulb increase is so slow that plants of twenty years' standing can still be vigorous yet not be overcrowded. (Pl. 18.)

L. nepalense while not unique is unusual in the way it sends its long shoots underground to appear at some distance from where the parent bulb was planted. It is one of the more difficult and rare species, but when seen in flower one is immediately aware of why it is in demand. The stems of two feet or more are furnished with glabrous leaves and finally carry two or three large blooms which seem too heavy to hold upright. They are really enormous, measuring six inches or more across. The colour of the outer lips of the reflexed petals is greenish-yellow, while the centre of the flower is filled by a large chocolate-red blotch. Greenhouse treatment is often advocated for this species, but the colonizing nature of its spreading shoots makes it a difficult subject for pot culture.

L. oxypetalum is another of those species so closely allied to Nomocharis that sometimes it is included in that genus. Plants twelve inches high carrying their single yellow blooms are the more welcome as they flower in late June. It is at present rare, although it is not a difficult plant, but building up stocks of this desirable yet difficult to obtain species can be a long-drawn-out affair. The search is well worth while. It is native to the north-west Himalaya. (Pl. VIc.)

L. *papilliferum*, an unusual species, is grown perhaps more by the collector than by anyone searching for colour. It gets its name from the papillae which protrude in quantity along the nectary furrow. These are brought into prominence by the fact that the dark blood-red flowers open widely, the petals being reflexed. Although discovered by the French missionary Père Delavay in 1888 it was not until after the last war that it flowered in this country. While not in the fore-front of garden plants its odd appearance has singled it out and bulbs are offered for sale.

L. *pumilum* is not the smallest lily, although its specific name when translated means dwarf. The flowers are not large, however, neither are its leaves, its foliage being linear and scattered around the wiry stem. This grows up to two feet and bears bright scarlet Turk's cap flowers early in the season. There is also a form bearing yellow flowers. L. *pumilum* is sometimes offered under its synonym L. *tenuifolium*, a name which more accurately describes the slender nature of its foliage. It is not of long duration in gardens, but may be raised from seed.

L. *regale* with its delightfully scented flowers is the lily which appeals to everyone. It was discovered by Ernest Wilson in Szechuan in 1903 and is quick to bloom from seed. Although bulbs may last for some years it is advisable to raise fresh stocks regularly. The flowers are white, flushed rose-purple on the outside. They are produced in a head and are funnel-shaped with splayed-out tips. The heavy scent is sometimes overpowering, especially within a confined space. Plants four feet high are quite handsome and are ideal for adding summer interest. The fringe of the peat garden would probably be the better site.

L. *rubellum* is another pink-flowered Japanese lily which is better planted in light shade to retain its delicate colour. It has broader leaves than most of the lilies so far discussed—especially L. *japonicum* with which it is sometimes classed—and it bears the same bluish meal when young. The trumpet-shaped flowers, however, do not open so widely nor are the plants as tall. In addition it flowers in early June, making it one of the first to bloom.

L. *sherriffiae* is mentioned only because it ought to be suitable for this situation, but its rarity and reluctance to settle in this country make more than a few words superfluous. When it can be encouraged to grow—and one or two of our best cultivators have managed to keep the plant alive—its purplish-brown flowers, speckled in the same way as *Fritillaria meleagris*, make it unique among lilies. The one-foot-high stems are invariably single flowered and, so far, only by continually raising new plants from seed can this species remain in gardens.

L. *taliense* was long a connoisseur's plant and was advertised as such in the catalogue of a well-known lily specialist. Its habitat in the Tali range of mountains in Yunnan was a favourite collecting area of George Forrest. Its recorded habit is similar to that of *L. lankongense*, except that in cultivation it never multiplies in the same way and used to be almost impossible to obtain. In recent years it has once more become available, but unless there is sufficient demand to warrant growing it the danger is that the nursery firm at present offering bulbs may cease to do so for economic reasons. It is best described as a white or pale-pink form of *L. wardii*.

L. tsingtauense is one of the easiest to grow. Whereas all the other lilies dealt with here either have pendent flowers, or ones which are held horizontal, this species has blooms which face upwards, displaying the shiny orange petals and stamens. Sometimes, on small stems, only a few flowers are borne, but on strong shoots numerous flowers can be packed in a head. Observant growers may also note that the foliage can appear distinctly mottled in the early weeks. Although said to be short lived, some colonies have been in existence for forty years.

L. wardii commemorates Frank Kingdon Ward, a very famous plant collector whose name will always be associated with plants introduced from the China-Bhutan border area. He made many trips in that region and sent back a large number of very fine garden-worthy plants. South-east Tibet is given as the distribution of this lovely lily. It is of the martagon type with petals completely reflexed. These are pink in colour and are heavily marked with small dark purple spots. Like one or two other species, *L. wardii* has stoloniferous stems which pop up at different places each year. There is no way of restricting this lily and it must be allowed to wander where it will.

L. willmottiae is probably more accurately placed as a variety of *L. davidii*. In many ways it is similar to that species, but it does differ in being not quite so tall; it appears to have many more leaves, which are certainly longer and drooping at their tips; and has flowers which are lighter orange in colour and are suspended on arching, drooping pedicels rather than being held out stiffly from the stem. The shoots, too, seem to emerge from the soil at an angle. This is an extremely handsome species or variety which must be included in any lily collection however small.

LIRIOPE is closely allied to Ophiopogon, a plant which was once much cultivated in pots and used to decorate the staging in conservatories. Liriope, on the other hand, is completely hardy and is extremely accommodating. One of its main assets is that it flowers in late autumn when most other plants are dying back. It forms tight tussocks of grass-like foliage, and the fact that it is evergreen is probably one of its lesser attractions, for the older exhausted foliage gives the clump an untidy appearance. Division in spring can produce many small offsets from a single divot.

L. muscari is found in China, Formosa and Japan and is considered a synonym of *L. platyphylla*. However, it is as the former that it is better known. The flowers, although small, are bright purple and are produced in tight clusters of four or five. Many of these clusters are found on a spike, so that the whole forms a dense column of purple. It is not unusual for the head to flatten and become fasciated at the tip.

NOMOCHARIS can only be described in superlatives. So much pleasure is got from simply looking at one of its flowers that if the peat garden enthusiast had to confine himself to twelve genera, Nomocharis would not be last on that short list. All who know it want to grow it. It is very closely allied to the genus Lilium, in fact there are two or three species which, to put it unscientifically, could be placed in either. The conditions favouring Lilium are also ideal for Nomocharis, but whereas Lilium includes many species from all land masses in the northern hemisphere, Nomocharis, so far, has been found only in the higher meadows of the Himalaya, Burma and China. Its name, translated, means "Grace of the Pasture" and this could well nigh describe its position in the peat garden. There are a few species, some being available from nurserymen, so that the experimentalist need not feel frustrated. The tendency of the species to hybridize has been responsible for the race of untypical plants often seen, and while the purist may scorn such progeny when searching for true species he can never deny their beauty (Pls. 28 and 29). They bloom in June and July. As the bulbs cannot be relied on to increase, seed is the only way by which stocks can be enlarged.

N. aperta is usually seen as a plant two feet high. The typical lily-like stems have leaves which are fairly broad and are topped by four or five flowers on long pedicels, which carry them well clear of the stems. They are wide open, the stamens prominently exserted, while the petals are a shade of rosy purple. The broad but not too intensive spotting round the centre of the flowers is typical of the species.

N. farreri is often considered a variety of N. pardanthina. It is finer than that species, both in the long narrow shape of its petals and in its whorled foliage. Strong plants may reach three feet in height, be perennial and be virtually trouble free. A vigorous clump of this species has, apart from one move, been continuously grown at the Royal Botanic Garden, Edinburgh, for forty years without ever relying on seed. Numerous flowers arise from the axils of the upper leaves. These are delicate pink with the central part of the inner segments heavily spotted with fine dots. (Pl. Xc.)

N. mairei is much in demand as a flowering plant. The flowers are flushed pink, but it is the heavy spotting of the petals, particularly the broad, fringed inner segments, which calls for attention. This is very dense and is widespread over the whole surface of the three inner petals. (Pl. Xa.)

N. nana is probably the least attractive so far as flower colour is concerned, but it is vitally important to a collector. It has been in cultivation a long time, in fact it was originally described under Fritillaria. Its drooping lilac to purple flowers are quite small and are carried singly on twelve-inch-high shoots. The leaves are long and narrow and the flowering season is June.

93

N. pardanthina is one of the loveliest members of this genus. The colour of the flowers can be almost white or pale pink, while the fringed inner segments suggest that this flower preens itself for display. Spotting adds to its attractiveness, but this is subdued, being confined to near the centre of the flower. In fact the dots in some forms are so zoned that they form a definite ring. (Pl. Xb.)

N. saluenensis is a little different from the others mentioned. It is more closely allied to *N. aperta*, but has broader leaves, a green zone in the centre of the flowers and an ovary which is much more prominent. Its petals are rosy purple, all lightly spotted with fine dots, but it is the greenish area which is so pronounced. This is a truly beautiful plant, but it would be unfair to rate it higher or lower than most other species. (Pl. Xd.)

NOTHOLIRION is said to be composed of four species. Not all are easily managed, in fact their winter's growth can be damaged to the detriment of flowering. The bulbs have the habit of dying after flowering, but so many small bulbils are spawned round the remains of the dead parent bulb that it is rarely lost. Growing on these small bulbs to flowering size is just one of the techniques of cultivation. The straight, tall flowering stems of the more vigorous species should be staked so that the long spikes are held proudly erect, and this also serves to preserve the seed capsules until they are harvested.

N. bulbuliferum has lavender-coloured flowers and, like the following species, the petal tips are green. Numerous flowers are borne on each stem. This plant was known for a long time under its synonym *N. hyacinthinum*.

N. campanulatum has crimson bells for flowers, but the tip of each petal is bright green. This combination makes it easy to identify. The flowers can measure up to two inches in length and are carried singly in the axils of the upper leaves. Four-foot-high stems are not exceptional.

N. macrophyllum, at one time known as *Fritillaria macrophylla* because of its much shorter growth, twelve inches or thereby, does not produce as many flowers as others do. They have a delicate colouring, however, being pink with a lavender tinge. Growing high in the hills of Nepal, Bhutan and Tibet, it should be hardy, but an early spring followed by late frosts can cause damage.

N. thomsonianum was once called the rose-coloured Lily. At that time it was labelled *Lilium roseum*. These two names describe the flower colour, a pale rose, and if flowering spikes are produced these appear in May. It is probably the loveliest of this fascinating genus despite being difficult to manage. The trumpet-shaped flowers also emit a delightful scent. Flowering spikes three feet high are the norm.

OAKESIELLA, often included in Uvularia, is a North American woodland plant which revels in a partially shaded site in soil which is loose and spongy. There it will send out its white, pink-tinged underground stems which may travel up to twelve inches or more before appearing above the ground. It is an attractive plant which quickly forms a dense mat of shoots and, while this may be advantageous in an area given over to self-managing plants, it can be a problem among the miniatures. The siting of this plant, then, is important, but the peat garden environs may have a home for it.

O. *sessilifolia* has, as the name suggests, sessile leaves. These are spear-shaped and at each node where they join the stem it changes direction so that a zigzag pattern is given to the shoots. These are usually more than a foot in height, and one to three yellow bells appear in the axils of the upper leaves. A form with variegated foliage is even more attractive, as the white markings brighten up that place in the shade which has been chosen for it.

PARIS, of course, is well known to the native flora enthusiast, and the indigenous species is always a thrilling find when botanizing. It is affectionately known as "Herb Paris" or "True Love" or, to give it its proper name, *P. quadrifolia*, and is one of our fascinating woodland species. Still another species, this time native to Turkey, is *P. incompleta*, introduced recently, but doubtfully still in cultivation. To class such plants as garden-worthy in a decorative sense, however, is questionable, but there is one which warrants that description.

P. *polyphylla* is native to the Himalaya. It is slow to emerge in spring and one must beware losing patience and damaging the young shoots by searching for them. It flowers in summer, the single flowers developing above a unique whorl of leaves, strong stems being up to three feet in height. The large seed box is extremely obvious with protruding styles, while the yellow petals, reduced to long but extremely thin strips, can be almost overlooked as floral parts. They remain attached to the lower side of the ovary the whole time. The plants are rhizomatous and after a few years may be divided, but if seeds are sown straight away, although they are slow to germinate, stocks can be raised more quickly. George Forrest sent home seeds on more than one occasion, and it must be said that these are not difficult to see. They are bright orange red and are revealed when the seed capsule splits open. (Pl. VIIc.)

PHILESIA is classed as a shrub. It is certainly evergreen and produces suckers which sprawl over and through the soil. Slow to become established, it is not a plant for every garden as our severe winters can be unkind to it. In its native Chile it is a woodland species and a moist shaded part in our gardens will bring forth the best results.

P. *magellanica* is the sole species and has dark green, lanceolate, curved leaves. The plant's form is a twisted entangled mat, and a position at the base of one of the walls is perhaps the best place to start it off. From then on it will make its own way by suckering shoots. It is unfortunate that the flowers are hidden within the foliage as these are large, waxy, translucent trumpets, two inches in length and bright crimson.

POLYGONATUM immediately brings to mind "Solomon's Seal", a very useful plant in any garden. The polygonatums have creeping rhizomes but are not really invasive. They are completely hardy, and once established need very little attention and in most cases they provide graceful lines and pale attractive foliage.

P. *hookeri* is an extremely dwarf species, little more than an inch high, and forms a mat of rhizomes and foliage. The solitary tubular lilac flowers are carried in the leaf axils and in packed colonies many go unseen. It spreads slowly, the underground shoots may be separated in spring, but the top growth quickly dies down in summer.

P. *japonicum* aesthetically is a most attractive species, especially the variegated form of it. The pale green foliage is broadly obtuse with the deep furrowed veins clearly seen. These are arranged alternately on gracefully arching shoots. The undersides of the leaves are distinctly glaucous, and sheltering there are the clusters of flowers. These are white at their tubular bases while the petals are pale green. Leaves which are pink-flushed with cream-coloured margins denote the cultivar 'Variegatum'.

P. *multiflorum* is the ubiquitous "Solomon's Seal" and, for a woodland plant of grace and form which gardeners have learned to appreciate, few plants can excel. The perfect line of its arching branches, the few-flowered clusters of white blooms which appear suspended beneath the foliage and the pale green of the leaves make a most attractive combination. It is trouble-free and revels in partial shade. There is a form with double flowers, but it in no way supersedes this wild European species which forms a picture when framed between the leaves of tall rhododendrons. It is one of the parents of the garden-worthy P. × *hybridum*, the plant frequently met with in cultivation.

P. *verticillatum* is so different that collectors of liliaceous genera must own this. The feature here is that the leaves are produced in whorls as opposed to being alternate. They are long and narrow, curling towards their tips, and as this means a number of leaves join the erect stem at the same place, so, too, a number of bunches of small greenish flowers arise from the axils. The tiered effect of the foliage is even more pronounced when the plants are in bloom.

RUSCUS is the genus to which the well-known "Butcher's Broom" belongs. It is an unusual evergreen plant in that the green, flattened, leaf-like structures are in fact modified stems. They are known as cladodes and perform the same functions as do leaves in other plants. These are present, nevertheless, and are to be seen as quite tiny structures in the middle of these cladodes. The flowers appear at that same spot and later, if fertilization has been successful, red fruits will follow to give colour in winter. These plants need an annual tidy up as whole shoots die back to be replaced by new suckering growths.

R. *aculeatus* is the common one and is ideally suited to a dark shaded area. It almost furnishes the impossible, and if there were a sunless area in the vicinity of the peat garden this species would help to clothe it. Stems three feet high are common. A form carrying narrow leaves and labelled *R. aculeatus* var. *angusti-folius* provides variation.

R. hypoglossum is not quite so amenable to any poor site, but its foliage is so broad and its young stems and cladodes are so polished that one has no wish to ban it to the background. The cladodes can be four inches by two inches, the stems up to twelve inches in length and the small true leaves are on the upper sides of the cladodes.

R. hypophyllum is so named as the tiny leaves are to be seen on the undersides of the cladodes. It is not as robust as the previous species, being barely nine inches tall, but if unusual plants appeal to the garden owner it must surely be classed in that category.

SCOLIOPUS is one of those miniature rarities sought after by alpine plantsmen. It is a completely hardy Californian native which apparently grows on moist shady slopes and in redwood forests. The peat garden is an ideal home in this country. Although closely allied to the beautiful "Wake Robin", the individual flowering stems of Scoliopus barely exceed an inch. Division as for Trillium is one way of increasing stock as seed does not appear to form in cultivation.

S. bigelovii, named in honour of Dr John M. Bigelow, an amateur botanist, has tiny upward-facing flowers with petals which are striped green and reddish-purple. The styles, divided into three, project skywards and appear relatively large. March is the month when, from fat, scarcely open bud tips poking through the ground, the small flowers emerge. Later the foliage lengthens, and although only a few inches long tends to look coarse when the smallness of the flowers is remembered.

SMILACINA is composed of more than twenty species, all of which are confined to the north temperate regions. On the whole they are native to woodlands and, in common with many other shade-tolerant plants, rely on underground stems to infiltrate an area. Some spread more quickly than others, but one or two have quite short rhizomes. None is stupendously floriferous, but they tend to have light foliage and white flowers, providing a combination to brighten shady corners.

S. racemosa has a proud erect habit. Twelve or more broadly spear-shaped leaves, pale green and soft to touch, amply furnish the three-feet-high stem. On top of this the six-inch-long many-flowered multiple raceme is carried. The flowers are cream coloured at first, but are pure white by the time they are fully open. From early spring when the reddish-brown sheathed buds burst through the soil until the herbaceous stems die back, this species has an architectural appeal.

S. stellata is, like the previous species, a North American native plant fairly widespread on that continent. It is less spectacular than *S. racemosa*, being little more than twelve inches high and carrying at the tips of its shoots short racemes of pale cream-coloured flowers. Although sparse, this species imparts a lightness to a shaded corner which might otherwise be dull.

SPEIRANTHA is monotypic and was collected in China not far from the international port of Shanghai. It is not reliably hardy in every garden and would probably be difficult in more inland colder areas, but at Edinburgh it is grown out of doors and has flowered for many years. A cool greenhouse is sometimes recommended, but if such a structure is not available one should not put aside all thoughts of growing it.

S. *gardenii* was at one time considered so like "Lily-of-the-Valley" that it was named S. *convallarioides*. Superficially this may be so, but actually the leaves in this case are evergreen and thicker than those of Convallaria, while the white flowers, although borne in a four-inch-high spike, are open star-shaped and not pendulous bells as in our own native plant. It spreads much less quickly, but differs in having flattened underground shoots.

STREPTOPUS is a small herbaceous genus, but in it there are at least two species well worth garden room. Uvularia and Disporum are two genera with which they were at one time confused. Streptopus means twisted foot or stalk and is said to describe the twisted flower stalk, a characteristic not readily seen. One could not be blamed for thinking that it probably refers to the zigzag pattern of the stem.

S. *roseus* is a hardy plant which occurs in a wide band across North America in the latitude of Newfoundland. It appears in an early issue of the *Botanical Magazine* as *Uvularia rosea* and is cited as having been discovered around 1806. It occurs in moist woodland in conditions similar to those of the peat garden. The ovate-oblong leaves are fairly heavy when compared with the small rose-coloured bells.

S. *simplex* is said to require greenhouse treatment, but once again this is a species which has shown its hardiness by growing strongly and flowering out of doors. This delicate Nepalese species produces arching shoots two to three feet long and it is in the axils of the upper oblong leaves that the tiny graceful bells develop. They are suspended on long slim pedicels which allow the white, pink-flushed blooms to bounce and sway in the least breeze. There is something fairy-like in the way this plant displays its flowers. The leaves possess a slight bloom in spring and are further enhanced by the redness of the young shoots. These pierce the soil with sharp points and quickly elongate until flowering starts in May or June. (Pl. 64.)

TOFIELDIA could be passed over if it were not for one particular species. There are in fact upwards of twelve species native to the colder areas of the northern hemisphere and all may be increased by division in spring.

T. *calyculata* is central European in its distribution and forms a cluster of grass-like shoots from which arise the six-inch-high flowering shoots. The very small, greenish-yellow flowers are packed into a dense spike and beyond the petals the yellow stamens protrude. A few square inches is all that is required to accommodate this very attractive dwarf plant.

TRICYRTIS has most peculiarly shaped flowers, the starfish-like stigma giving an odd appearance. The flowers are carried in terminal and axillary clusters and are upward facing. "Toad lily" is a name often used when referring to it. It is perfectly hardy in Britain, but plants may be lost when transplanting. It is not that they are difficult to replant, but unless this is carried out in spring, just as growth starts, there is a danger that earlier divisions may be lifted by frosts and so fail to re-establish. They are valuable autumn-flowering plants.

T. *formosana* has slender stems which are almost devoid of hairs and can reach eighteen inches in height. They carry sheathing, lanceolate leaves and purple-spotted lilac flowers. The species is apparently widespread in Taiwan, where it grows at an altitude of around 7,000 feet.

T. *hirta* is rather common in Japan. It is pubescent, in fact the white form is densely covered with white hairs. Growing to a height of three feet it is prominent and displays well its large sky-facing flowers which have a white background and are heavily splattered with purple spots. T. *hirta* 'Alba' has pure white flowers.

T. *latifolia* is worth including in the collection because of its yellow blooms. Even these are purple-spotted, but the yellow colouring is dominant and when in flower makes this Japanese species conspicuous. Even the stems are prominently speckled.

TRILLIUM, "Wake Robin", "Lamb's Hindquarters", "Trinity Flower", "Wood Lily" are all names affectionately applied to various species, and what a fascinating group of plants this is. From the moment the first shoots start developing in early spring until the last species has bloomed there is an exciting cavalcade of types. Trilliums must be among the easiest of woodlanders, and from experience they appear to be trouble free. Their biggest enemy is probably man himself when he neglects them or, during their less than dormant season, inadvertently digs them up or treads on their crowns. It may be just as well to draw attention to this apparent lack of resting period, for even in December their new shoots are well developed and piercing the soil. They are also the most amenable of plants to transplant. In autumn—October—clumps may be lifted, shaken out completely and, after the various pieces of rhizome have been sorted out, they should be replanted, giving each portion room to multiply. Some species are slow to increase in this way, but eventually, when side shoots have obviously developed, one need not fear handling the clump. Seed forms in some instances and this can be used to increase plant numbers, but germination is slow and it may take more than one year for all the seeds to grow. Growth in the early stages, too, is slow, so that one must just be patient. After all, this is one of the most satisfying things about gardening, the growing and increasing of one's own plants. There are odd plants, however, which are obviously clones and appear to be self sterile and certainly fail to set seed. They also seem unable to increase vegetatively, and in one instance (and this refers to two identical plants received in 1947), although they grow strongly, are healthy and vigorous and flower every year, the roll call would still be answered by two rhizomes. Their distribution ranges over practically the whole of North America, Japan, parts of China and the Himalaya. As with Rhododendron, Trillium and peat gardening are almost synonymous.

T. cernuum is an easily identified species producing an umbrella consisting of three almost round leaves. The bent-over flower stalk carries the bloom beneath this canopy and on examination these flowers will be seen to be quite large. Three very much reflexed petals display the stamens and the prominent, deep pink seed box. This plant comes from eastern North America and has a stem measuring upwards of twelve inches. (Pl. 65.)

T. chloropetalum has often been confused with *T. sessile,* a mix-up which is likely to remain for a very long time. It is, however, a much larger plant in every way. Both are from North America, but *T. chloropetalum* is from California while *T. sessile* is from the middle and eastern states. The species under discussion is a most spectacular plant and, while its flower colour may vary from red to pure white, in all its forms the erect petals sit tightly in the centre of the three sessile leaves. It is fascinating to watch the white or red petals emerge from the three sheathing bright green sepals. Finally this plant may reach eighteen inches in height. (Pl. XVIc.)

T. erectum is the species referred to as "Lamb's Hindquarters". Using imagination one might be able to liken the red flowers to that part of a lamb hanging in a butcher's shop. The collar of three leaves usually forms twelve inches above the soil and on top of this the blood-red flowers, held on a short stem, arch over until they are virtually at right angles. Single plants may not be too impressive, but a small group can cause a stir (Pl. 66). White forms of this species occur; one well worth searching for is *T. erectum* var. *albiflorum.* If marks were given for a blending of greens and whites this variety would lose none (Pl. XVIb).

T. grandiflorum flowers during April and May and is probably the species most often seen. It has a well-balanced arrangement of vein-furrowed leaves and a pure white large flower. This is funnel-shaped, but even when fully open the edges of the petals continue to overlap (Pl. 67). Semi-shaded sites provide the right conditions and no plant is easier to manage. Beautiful as this species is, a pink form can be had which has petals of the most delicate shade. The foliage, too, has an extension of this pink to purple flush and a plant of *T. grandiflorum* 'Roseum' is a real treasure (Pl. XVIa). Double flowers, generally, are not usually considered improvements on single flowers, but *T. grandiflorum* 'Flore Plenum' (Pl. 68), a form found in the wild, is a collector's item. It is not too plentiful as yet, but its rate of increase suggests that more people may have the pleasure of growing it soon. The multiplication of pure white petals increases the flower's substance, and one could not be blamed for comparing it favourably with the Gardenia.

T. luteum is yet another species with stalkless flowers, but, as the name suggests, the colour of the petals here is yellow. It is not as robust as either *T. chloropetalum* or *T. sessile*, being usually nine inches or less in gardens, but it has the attractive mottling of the leaves so prominent in this group.

T. nervosum is a slender, delicate-looking species from south-east United States. It has wide nodding flowers, light pink in colour with petals which are much reflexed. The fact that the flowers are below the level of the leaves matters little as the leaves, too, are thin in texture and have wide gaps between them so that the flowers are visible. *T. catesbaei* and *T. stylosum* are synonyms, and the former is even regarded at the present time as the authentic specific name by some botanists.

T. nivale, the "Snow Trillium", is one of the smallest species. It barely reaches five inches, yet has quite large flowers for so dwarf a plant. The petals are pure white. Because of its lack of stature it needs careful placing, i.e. among autumn gentians or some other ideal ground-covering plant so that they can form an association. Cultivation and therefore careful maintenance are then more certain. This is a plant for the connoisseur.

T. ovatum is one of the earliest to bloom, the first flowers, depending on the season, being open by the end of March. The buds expand soon after the shoots appear above ground, showing the partially opened flowers nestling in the, as yet, only slightly developed leaves. All quickly enlarge until the plants measure almost a foot in height. The flowers tend to be held erect, and although flowering early this is one species which invariably sets ample seed. In fact seeds which fall and are left undisturbed can form colonies over the years. It is not a refined species and the foliage tends to be coarse, while the stout stalks, reddish in colour, are not unlike miniature rhubarb stalks. *T. ovatum* 'Kenmore' is a double form. So far it has shown little tendency to increase, but as it falls far short of the double *T. grandiflorum* it is no aesthetic loss.

T. rivale, known as the "Brook Trillium", belongs to the Californian flora. It can best be described as a pink mottled counterpart of *T. nivale*. The flower stalk is also much longer, while the leaves, too, seem to be more widely spaced. This is one of the peat garden's treasures. (Pl. 69.)

T. sessile has stalkless leaves and flowers. The broad marbled foliage resembles a close collar. On top of this the sepals lie across the area where the leaves overlap, while the inner floral parts, as in *T. chloropetalum*, remain stiffly erect. The colour varies from translucent crimson to dark maroon. (Pl. 70.)

T. undulatum may be the last of a line of species, but it is one of the most distinguished. While spotting of the flowers may have been significant in one species, here the name "Painted Trillium" records that the flowers are shaded. Held above the foliage they are white, flushed or streaked pink with a heavy fringed zone of magenta in the middle.

To make a selection from the species listed would be difficult as all are aristocrats of the plant world, but, if one were restricted to a single species, *T. grandiflorum* could hardly be ignored.

UVULARIA is a very small genus which contributes well to interest in the garden. It is herbaceous and in May has drooping, long-petalled, yellow flowers. The distinctive foliage remains for the rest of the season. It spreads slowly, the clump increasing in size each year, and may be divided and re-planted in March when growth commences.

U. grandiflora always appears to be suffering from drought. Even the bright yellow flowers suspended at the ends of the twelve-inch-high stems have peculiarly twisted petals. The upper leaves, too, remain partly rolled during the time the plants are in bloom, a characteristic which is somewhat unfortunate. (Pl. 71.)

U. perfoliata is very closely allied to the above, but is not quite so robust. The flowers are slightly smaller and while the leaves, too, are slightly rolled they seem to assume their normal flat plane more quickly. This makes the glaucous bloom, pronounced on the undersides, easy to see.

XEROPHYLLUM, a name which refers to the dry grass-like leaves, hardly describes the dense ray-like arrangement of the individual crowns. The tough evergreen leaves, up to six inches long, have a grey-green appearance and a ser-rated edge. This is not a plant for shade, but it will appreciate a moist peaty soil and a chosen site.

X. asphodeloides is handsome enough in the rosette stage, but when it flowers it can send up a spike three feet in height, well furnished with hard narrow leaves topped by a dense head of creamy-white blooms. Many flowers go to make up the inflorescence and, as the topmost ones develop, the flower head becomes a perfect dome. The division of the crowns in spring and seed sowing will ensure new plants.

10 FROM OTHER PLANT FAMILIES
(*Amaryllidaceae to Umbelliferae*)

ADONIS (Ranunculaceae) is named after the Greek god of that name who, in keeping with the botanical naming of plants, was considered the type in manly beauty. The plant christened after him certainly belongs to a class of beautiful plants. There are a number of species native to the northern hemisphere. Some are perennial and flower early, emerging into the upper world for six months as Adonis did in spring. The annuals, which flower in summer and autumn, although attractive in their own right, are not considered here. Perennial species, once in place, prefer to be left alone. They form a woody rootstock, a bunch of attached crowns, which are difficult to separate; in fact pieces broken off, even although they are later grown in pots for a year, are slow to re-establish and may sulk for a few years. The roots which feed the plant penetrate deeply into the soil so that the retention of young feeding roots is virtually impossible. Raising plants from seed, though germination is slow and erratic, is the most satisfactory way.

A. amurensis, although first collected on the banks of the Amur river, an important watercourse on the Chinese-Russian border, has since been collected in Korea and the larger islands of Japan. Its flowers, which are single, are composed of an uncertain number of petals—twenty to thirty—and display the typical glossy surface when they are fully open. The reverse sides of the petals are streaked with green while the three or four which initially formed the bud have bronze shading. The blooms are carried on nine- to twelve-inch-high stems which are themselves adorned with deeply cut fern-like foliage. While the yellow flowers are more usual there are plants which show the species variation by having flowers which are pink, cream or even striped, and occasionally some are imported from Japan. A double-flowered variety is perhaps the one most often seen. Labelled *A. amurensis* 'Flore Pleno', the boss of the flowers is usually a mass of tiny green petals.

A. brevistyla, a Himalayan species, brings a new flower colour to the genus. Here it is pure white, the delicate petals surrounding a cluster of golden yellow stamens. Before opening, and in dull weather when, like the daisy, the flowers remain closed, the blue stripes and shading are visible. Flowers open when the shoots are barely six inches in height, but the stems continue to lengthen until, at fruiting time, they can be up to fifteen inches. The divided foliage is an additional feature.

A. vernalis should be in everyone's garden. Although wild in central Europe it is one of the loveliest plants cultivated. It flowers in April, and, when the sun shines on the four-inch-wide yellow flowers, dazzling is the only word that can describe the effect. The reflection from the polished surfaces of the twenty or so petals adds to the brightness. When closed, the outsides of the petals are polished copper. The solitary flowers are borne on six- to nine-inch stems and these carry the numerous finely dissected leaves. Large fruiting stems, eighteen inches tall, finish the season. (Pl. 1.)

ANEMONOPSIS (Ranunculaceae) is monotypic and, while this state occurs in other parts of the world, Japan seems to have more than her fair share of those that are garden-worthy. The name means Anemone-like, which seems to describe quite accurately the plant under discussion. Even in nature it is apparently quite rare, frequenting wooded slopes on some mountains in central Honshu. Fortunately it is not scarce in cultivation and enjoys a well-drained peaty soil and can be increased by seeds or division in spring.

A. macrophylla will reach three feet in height, but, as the divided leaves are virtually confined to the lower third, and the upper branching flowering stems are well spaced and carry smallish flowers, no shade or suppression is effected by it. An inspection of the individual flowers is most enlightening. The three outer sepals of the bud are greyish-purple on the outside, while the faces of the inner ones, eight or nine in number, are white with just a suggestion of blue. In the centre and standing erect, as though protecting the ovary and stamens, is a ring of approximately twelve petals. They almost form a closed ball and have purple-shaded tips. The unusual inverted follicules (seed capsules) at fruiting time should appeal to the floral artist.

ARISAEMA (Araceae) is a large genus belonging to a very big family of which only a few are of interest here. Many are tender, but among those that are hardy can be found a few which are bizarre. Their rootstock is a depressed tuber, but their elongated growth, consisting of one or two leaves and a hooded spadix (flower spike), provides the unusual. The tubers should be planted two inches deep in a vegetable soil, after which it is important that they receive plenty of moisture during the growing season.

A. candidissimum was discovered and collected by George Forrest in Yunnan, and later by Frank Kingdon Ward who described it as being a member of the herbaceous flora of the pine forests. Its roots go deep in search of moisture and so its place in the peat garden, as with all other aroids, is probably better in a drier part. Everything depends on the local rainfall and, of course, the drainage. The single trifoliate leaf, bright green in colour and carried on a long petiole, is complementary to the large, handsome spathe. This is green striped with white at the base, while the hooded part is white flushed with pink.

A. ringens is one of many species native to Japan. If the situation suits this plant it will flower annually and be reliably perennial. The basic colour may be greenish or purple, but lined on the outside with numerous green stripes. These travel up the tube and over the top of the hood. The lip of the spathe and the appendage to the hood are dark purple. A suitable site is in among dwarf rhododendrons or with a ground cover of autumn flowering gentians. Propagation is by little side tubers.

A. triphyllum is North American, but self-sown seedlings, which can appear round the bases of rhododendrons and where the turf meets the soil, indicate how much this species could spread if cultivation did not disturb its progress. The heavily spotted leaf stalk is quite interesting, but it is the green spathe striped with white and brownish lines which causes most comment. The green hood resembles the flap of a purse in the way it bends over the top of the spadix, protecting it. A form listed as *A. triphyllum* var. *zebrinum*, illustrated in the *Botanical Magazine*, plate 952, has a white-striped purple spathe which would be worth procuring if it were ever available.

ARISARUM (Araceae) contains very few species, in fact the genus is confined to the Mediterranean basin, but only one species is recommended here. This can be grown in quite dense shade, a factor that makes it useful for clothing dark corners. It also tolerates full sunshine and flowers quite well there, but the leaves in shade always looks much better, being darker green and larger. The tubers are slim and tiny and only require to be an inch below the surface.

A. proboscideum gets its name from the proboscis or long thin snout which is a continuation of the spathe. This alone measures four inches in length and is attached to the end of the brownish-purple hood. Beneath this and towards the base there is a definite demarcation line after which the spathe becomes much lighter and in part is white. The leaves, less than six inches long including the stalk, are auriculate, that is to say that in their shape they include two prominent ear-like additions close to where the leaf joins the stalk. Usually the foliage is so dense that many of the flowers are hidden beneath its canopy. The "Mouse-tail" plant is a common name by which this species is identified and this is appropriate because it is the long mouse-tails which are so prominent in May. (Pl. 3.)

ARUM (Araceae) is yet another genus which encloses its flower spike within an inflated spathe. This most spectacular way of drawing attention to flowers is epitomized in the "Lily of the Nile" or "Arum Lily" of the florists. Nothing quite so grand is intended here nor would anything so exotic or gross be in keeping with the usual peat garden plants. Many species are indigenous to the countries bordering the Mediterranean and there are two which are well dispersed throughout parts of the British Isles.

A. italicum, although the specific name suggests Italy as its place of origin, is widespread throughout Europe, including a number of localities in south-west England and Wales. Although herbaceous in character its leaves appear in winter and remain fresh until the summer. As the white markings, so prevalent on the leaves of this species, are one of its main attractions it is obvious that they will last longer in a reasonably moist site. The spathe is pale green while the spadix is yellow. The name *A. italicum* 'Pictum' is a clonal one sometimes given to forms which show extensive white patterns.

A. maculatum is our native "Lords and Ladies" and "Cuckoo-Pint". This species, well known to all who study British flora even casually, is quite at home in the garden. The spotted spathe and foliage are most decorative, and these are followed in autumn by equally attractive clusters of bright red berries bare of all foliage and standing six inches high. This is a plant well worth considering for one of the many shaded sites under shrubs.

ASARUM (Aristolochiaceae), known in North America as "Wild Ginger", is a peculiar plant with creeping stems. Like Arisarum, it does not display its blooms prominently. The flowers of some species lie on the surface of the soil while others are upright, but in all cases they are hidden among the leaves. In nature they seek the shade of trees, being inhabitants of redwood and pine forests in North America and various plantations in Europe.

A. caudatum, the specific name meaning 'having a tail', describes the three 'tails' which form part of this flower. The flower has three lobes and the tips of these are projected into these flagellate appendages. The petals are reddish-brown, lying flat on the ground within the shade of the broad heart-shaped leaves, which form a complete ground cover.

A. europaeum, our native species, does not spread so quickly. Clumps may be divided in spring and quite small plants can yield many young pieces suitable for replanting. The foliage in this case is round to kidney shaped and dark glossy green on top. The dull brown flower—the perianth of which never opens out completely—is carried on short thick stems and hangs from between a pair of leaves.

ASTILBE (Saxifragaceae) is usually associated with ponds, moist areas and herbaceous borders. While being colourful and gay and first-rate garden plants, some are much too tall and garish and would be incongruous here. So it is fortunate that smaller species are available which can introduce featheriness and lightness to the peat garden as well as continuing the flowering interest into late summer and autumn. Just like their larger allies they require ample moisture in summer to keep them growing and, because they are small and some have shallow roots, top dressing in spring helps in acting as a protecting mulch. Regular division and replanting are also beneficial, as this avoids having clumps which are woody in the centre where only small weakly leaves are produced.

A. chinensis 'Pumila' can add colour in a moist shaded corner from August onwards. Each tight erect flower spike, up to fifteen inches tall, carries thousands of tiny purplish-red flowers. The spike is composed of a large number of closely packed racemes. Even the foliage is handsome, being extremely indented while the margin seems to be defined in red. The species itself, a native of China, has a tall flower spike up to three feet in height and does not look quite so well balanced.

A. × crispa is a grex name, that is a collective name given to a swarm of hybrid forms raised by crossing two species and before any kind of selection has taken place. From this variable collection cultivars bearing fancy names are chosen, examples being 'Lilliput' and 'Gnome', which clearly give an idea of their stature. Here they are barely nine inches in height. Their dark crinkly almost furze-like foliage is massed just above the soil surface and, prior to flowering, resembles a dark green miniature fern. The densely packed flower spikes arise in midsummer and remain attractive into autumn. Ample moisture, rich soil and frequent replanting are essential to keep the plants growing.

A. glaberrima var. *saxatilis* should be included in any collection of dwarf hardy plants. This tiny individual, scarcely four inches high and decorating the soil with firm and deeply cut bronze foliage, needs protection from more vigorous plants. Its pigmy inflorescences are dotted with infinitesimal pink stars.

A. simplicifolia is native to Japan. Its only station is on the island of Honshu and even there it is classed as rare. This delightful species has most elegant arching branches with heights ranging from three to eight inches. On these the inflorescences are borne in large numbers. They seem to billow out and from a distance resemble white foam. Unlike other astilbes this species has simple leaves, although the edges are frilled, or, to be more explicit, are doubly toothed. Slightly larger coloured forms are sometimes offered, but these are in fact hybrids, *A. simplicifolia* being one of the parents.

BOYKINIA (Saxifragaceae) to a rock garden enthusiast immediately brings to mind the species from Pike's Peak, Colorado, *B. jamesii*, but that is not a peat garden plant. Named in honour of Dr Samuel Boykin, an American field botanist, the genus includes a few which are woodlanders. These are all perennial and, on the whole, hardy. Only two are mentioned here, for it is not a large genus.

B. occidentalis is a shallow-rooted North American plant, forming mats of vegetation which can, after a short number of years, become less floriferous. It is therefore important to divide it frequently to ensure flowers. While the stems, leaves and calyces are reddish in colour the small flowers, carried in loose panicles, are white. The leaves are deeply lobed and serrated and the crown of the plant is somewhat hairy. This species has a lightness not often found in plants which are shade-tolerant. It closely resembles a woodland saxifrage.

B. *tellimoides* is a more robust plant than the previous one, having deeply lobed leaves, the lobes themselves being coarsely toothed. The leaf-stalk joins the blades about the middle and it is from there that the prominent veins radiate. It produces an umbrella-like effect. The underside of the foliage has a glaucous hue. The basal leaves are quite large and may measure five or six inches in diameter. Two feet is the maximum height of the racemes and, if one looks into the creamy-white frilled flowers which develop in summer, one will notice ten dark purple anthers forming an inner ring round the stigma. This species was first placed in Saxifraga before its affinity to the North American plant had been noticed, but if one wished to be even more up to date the generic name could now read Peltoboykinia as accepted in the most recent Japanese floras.

CALANTHE (Orchidaceae) is a wonderful genus containing many species of high merit. Few, if any, could be termed hardy, yet one need never be surprised at the amazing way in which tropical genera seem occasionally to have outliers which will tolerate outside conditions in a cold garden. In nature the environment preferred is a peaty, boggy one, very similar to the one being discussed, but of course it is the condition of the bog in winter, the dormant period, that is so vital.

C. *alpina* grows, flowers and flourishes outside in a garden as far north as Aberdeen. The site chosen is a north-facing border and its neighbours are dwarf rhododendrons, hostas and lilies. In summer the flowering stems rise twelve inches, the upper third being occupied by the well-spaced, yellow, orange and reddish-brown flower and sepals. Blooming so late it is the more highly prized, as so many of the other unusual treasures are spring flowering. The short rhizomes spread slowly through the soil. This species is of Himalayan origin. (Pl. III.)

CALCEOLARIA (Scrophulariaceae) may appear at first sight to be an odd type of plant to introduce into the peat garden. One's immediate reaction to the name is that the plants are not hardy; some may be half-hardy but on the whole they are for greenhouse floral decoration. Of the few worth considering, it can be said that they prefer a moist situation in dappled shade rather than full sun. In fact the tiny C. *tenella* would quickly dry out and disappear in light soil devoid of shade.

C. *biflora*, as the name implies, usually has only two flowers on each flowering shoot. These appear in summer, are pure yellow, of well-balanced proportions and are a half to three-quarters of an inch across the basal pouch. It is a good perennial forming tight low hummocks of glossy green leaves.

C. *tenella* carries an epithet which can mean tender and delicate, but in this instance refers to appearance. This little treasure, native to Chile, spreads by creeping across the surface of the soil, the stems rooting as they go. It flowers with abandon during June and July when hardly any of its tiny mat-forming foliage can be seen. The flowering stems are around three inches high and each carries a number of terminal and axillary flowers. If it can be said to have a fault it is that it blooms too well and tends to become exhausted. Frequent division and fresh soil are necessary. It was discovered in Chile in 1823 by the German traveller Poeppig.

CARDAMINE (Cruciferae) includes annuals, biennials and perennials, many of no horticultural value, some even being classed as weeds. One or two perennial types are worth considering and these invariably prefer cool, moist conditions in partial shade. They are completely hardy.

C. *asarifolia* is well named as the leaves are distinctly similar to those of Asarum. They are round with bases which are cordate, these lower edges filling in the space, in fact overlapping each other. It occurs in the south of France and in northern Italy and when first described in the *Botanical Magazine* was recorded as coming from Mont Cenis. The pure white flowers are produced in many-flowered racemes, these usually being about twelve inches high. It grows and flowers well in quite deep shade. (Pl. 4.)

C. *pratensis* is our own native "Lady's Smock" and, while a planting in a damp area by the side of a stream and perhaps overhung by tall trees could look colourful in May when the plants bloom, one would be reluctant to try to establish it in the garden. There is a form with double flowers, however, C. *pratensis* 'Flore Pleno', which spreads with reasonable ease and yet is not difficult to contain. Apart from its flowers being double it resembles the species in all other respects.

C. *trifolia* is a useful plant which slowly infiltrates soil bare of vegetation, for example the ground beneath rhododendrons or other shrubs which cast shade. The name indicates that there are three leaflets, each being toothed and almost round. They are dark green above and purplish on the undersides. The white flowers, carried on four- to six-inch-high leafless stems, appear in May, developing as the stems advance. This is a very fine addition to a collection of shade-tolerant plants.

CAULOPHYLLUM (Berberidaceae) is one of a number of herbaceous genera included in this plant family. Some of them will be dealt with in this book, for while none may be extravagantly garish they have a quality far above mere floral interest. They are singular and all have character.

C. *thalictroides* has small yellow flowers borne in a sparsely branching inflorescence appearing in May. Later, bluish so-called berries develop. Fifteen to eighteen inches is the usual height of this species in cultivation, although plants more than twice that height are apparently common in nature. It gets its name from its large divided leaf, there being only one which, without flowers, could almost pass as that of Thalictrum, the "Meadow Rue". It makes an ideal filler among the background rhododendrons.

CHIMAPHILA (Pyrolaceae) was at one time included in Pyrola, but that genus, while still used for some species, has been divided into one or two other genera. There are only a few involved here. All are a little difficult to establish in gardens but, as they are of necessity woodland plants and must have a lime-free soil, the peat garden presents the best possibility for success. They are dwarf sub-shrubs.

C. *japonica* is found not only in Japan; its distribution chart shows that it also occurs in Korea, Manchuria and China. The leaves are glabrous, broadly lanceolate, sharply toothed and thick to touch. Sometimes they are opposite, other times alternate and again they may be found in threes on the single or occasionally branched stems. One or sometimes two nodding white flowers should appear early in summer.

C. *umbellata* is also sub-shrubby and is native to all the northern temperate continents. The lanceolate toothed foliage is in whorls of threes. Projected beyond this are the small umbellate inflorescences, four or five flowers forming a circle on top of the pink stem. The rounded petals may be pink in the centre and cream or white on the outer half. The prominent style and stamens contribute towards an attractive plant.

CODONOPSIS (Campanulaceae) are held in high esteem by some growers, especially those who enjoy looking closely into the flowers. Some emit an unpleasant odour when any part is bruised, a factor which does not endear them. But there is no doubt that the pattern of veining and the shades of colouring that can be traced inside the open bells are most intricate and fascinating. Codonopsis are herbaceous climbers or scramblers and, while some may be so robust that they damage the plants over which they grope, quite a few have lighter growths which can clamber through a Rhododendron without causing it any harm. In the peat garden, as in the rock garden, if the tubers are planted on top of a peat bank, and the resulting plants are allowed to trail over and down, no harm should come to any neighbour and the flowers will still be displayed to advantage. Seed is the way to raise plants and should be sown thinly in spring and the seedlings pricked out into pots as soon as possible. After a year they may be planted out into their permanent quarters. Codonopsis have tubers which develop into contorted shapes and are not suitable for transplanting. As the young shoots are fine but vulnerable, a few twigs will help the twining stems on their first leg up to the lower branches of the intended supporting shrub. When weeding, care must be taken that these shoots remain undamaged.

C. *clematidea* is probably the worst culprit in so far as unpleasant odours are concerned. It is also a rank-growing species, the leafy growths spreading as much as four feet. Huge tubers are formed and old plants of forty years and more are commonplace. Numerous flowers are produced and seed is usually plentiful; seedlings which are self-sown can appear in the oddest places, but can be easily removed. The flowers are a pale shade of blue on the outside, but when one looks into the bells other colours are apparent, a definite pattern of yellow at the base brightening the whole flower.

C. *convolvulacea* entwines itself to a shrubby plant and climbs until its top growth emerges into the light. A position at the base of a Rhododendron is a suitable one. The leaves are long, lanceolate, sometimes up to two inches, but it is the large, wide-open, lovely blue flowers which are so distinguished. It is a pity so few flowers are open at any one time. It has a wide distribution extending through the Himalaya into western China.

C. *dicentrifolia* is easily identified because its foliage is similar to that of Dicentra. Its habit of growth is quite distinct from the last species, being more compact, up to a foot in height and spreading. The light purplish-blue bells are Campanula-like, the tips being curled back, but on the inside they are mottled with white. These bells can be quite plentiful and are comparatively large, at least one inch long.

C. *meleagris* has most unusual growth for a Codonopsis. In the early stages it forms a basal rosette of two or four leaves, and if the plant is of flowering size the shoots carrying the flowers will develop from these. The drooping flowers are axillary and terminal on these shoots, but what makes them so fascinating is the chequered pattern reminiscent of the "Snake's Head" Fritillaria. Prior to opening, the flower is distinctly pear-shaped and even after the bell is fully expanded there is a tendency for the tips to remain curled inwards. George Forrest discovered C. *meleagris* in grassy glades in pine forests in Yunnan in 1916. Its flowering season is midsummer.

C. *mollis* was first found by Colonel Walton of *Primula waltonii* fame. He found it on the hills above Lhasa in 1904. It is quite enlightening, when considering the different species, to note how diverse they can be. This one is extremely hairy with small ovate leaves. It produces flowering shoots up to eighteen inches in length on top of which the blue tubular flowers are borne. These are waisted about the middle while the blue is marked with reddish stripes on the inside of the tube.

C. *ovata* must be classed as one of the best and most decorative. It has a graceful habit, producing its flowers on strong twelve-inch-high shoots. These carry the wide-open bells and, while the shadings range from purple to light blue, the inside is much more decorative. There the ground colours are lilac to nearly white, and these are contrasted by the deep purple lines which trace the path of the veins. The whole plant is bristly with hairs, including both sides of the small ovate leaves.

C. *rotundifolia* is a climber like C. *convolvulacea*, except that the shoots are much more branching, resulting in many more flowering shoots developing. This species probably flowers as well as any and better than most. The flowers are not prominent, however; their colouring is yellowish-green, the veins being traced in red. Furthermore, unless one can look up into the flowering shoots, many are lost in the tangle. The insides, like those of most other species, show more intricate markings, and here a zone of small red flecks divides the corolla in two. The long to oval leaves are bright green, showing a decided bloom on the undersides.

C. tubulosa has flowers which are similar in shape to those of *C. mollis*, but here they are bright yellow with a purplish base and the whole plant lacks the pubescence of *C. mollis*. It is rare in cultivation and has been collected in Upper Burma and West China.

C. vinciflora is of much garden merit and, although sometimes compared with *C. convolvulacea*, is not as vigorous as that species. Its leaves, too, are ovate and small. If the tubers are placed among the roots of a Rhododendron, *R. russatum* for example, and a few twigs inserted round the emerging wiry shoots to encourage them to clamber up on to the branches, success is possible. Then one may enjoy the large bright blue flowers. The young growths are so brittle that any handling must be delicately done. It is to the intrepid plant collectors Frank Ludlow and Major George Sherriff that we owe the last introduction of this species.

CORNUS (Cornaceae) is a very useful and attractive genus. It is extremely variable, including some quite tall woody plants as well as others which are creeping and herbaceous. It is the dwarf species which spread by underground stems that are assessed here. Moist soils and even partial shade are suitable. The searching underground stems will go quickly through a peat block, riddling it and densely suckering until it is a mass of shoots.

C. canadensis, a North American bog plant, obviously enjoys a moist porous medium. It blooms in May and June, although flowers can appear spasmodically during the rest of the summer. What are often taken as flowers are really heads of small flowers, the four conspicuous white petal-like attendants being floral bracts. In some gardens red berries are produced in autumn. Even if no berries appeared, a garden which claimed to have a comprehensive collection of dwarf plants would be considered incomplete without this Cornus. Six inches is about the average height of this very desirable although spreading species. (Pl. 8.)

C. suecica is a dwarf herbaceous perennial widespread throughout the north temperate and subarctic zones. It is much less invasive in gardens than the previous species and less garden-worthy, but it is so interesting that if plants can be got attempts should be made to establish it. Open peaty soil suits this species, but there is no doubt it misses its associate plants when transferred to garden locations. Despite the wide natural distribution it will always be rare in cultivation. It is a plant readily distinguished by its oval, opposite leaves with three to five prominent veins which, apart from the central one, follow lengthwise the curve of the edges. A small cluster of flowers is attended by four conspicuous creamy-white bracts. These are apparent in early summer. As the two species listed here are very near relatives, the latter one has been retained under its synonym rather than its long accurate generic name, *Chamaepericly-menum*.

CORYDALIS (Papaveraceae) is a genus of shade-tolerant plants and is therefore suited to a peat garden environment. It is not necessary to plant in large drifts for mass display, but consider them as fillers, or embellishers, able to clothe areas of soil which might otherwise remain bare. Sited under the shrubs, or in the shade of taller perennials, they can greatly increase plant interest. Tubers of varying shapes and fleshy type roots describe the overwintering organs of most species although there are a few which bear fibrous roots.

C. cashmiriana is often considered the finest member, and some who have seen this species in flower may agree with this. However, there are other species, but none so far has a colour to compare with the bright sky-blue of the flowers exhibited here. It is a neat plant of barely five inches, carrying its small cluster of long-spurred flowers clear of the bright green divided foliage. In a few Scottish gardens self-sown seedlings appear, and in one particularly they are widespread in the shade of tall shrubs. The plant's main enemies are birds, for as the tiny tubers lie just below soil level the scratching of the passerines may disturb them, so that they can dry out if not noticed and replanted.

C. cava is European and taller than *C. solida*, a species with which it is often confused. It is one of the early flowering plants, being prominent from February to May, depending on the season, and producing purple blooms on single stems six inches high. It is a reliable perennial and of garden merit. A white form given the name of *C. cava* 'Albiflora' is sometimes available.

C. chaerophylla is quite tall and has a long flowering period. Its leafy stems may reach up to four feet, carrying much divided light green leaves and multi-flowered racemes of tiny golden yellow flowers. Later small seeds develop on the flower stalk and these can be reddish-bronze. It is an ideal plant for the background and, being native to the high Himalaya, is reliably hardy.

C. cheilanthifolia, a little gem, barely measures eight inches to the top of its flowering shoots. These are many and light yellow in colour, a shade which seems right to be associated with the delicate ferny foliage which more or less forms a ground cover for the flower spikes. This low-growing Chinese species should be planted near the front of the garden so that it may be enjoyed in all its beauty. (Pl. IIc.)

C. nobilis, truly handsome as its second name suggests, proudly carries its large yellow flowers in a dense head. The leaves are double pinnate and are produced in sufficient numbers to clothe ably the erect stems. Long-spurred flowers are typical of this species.

C. solida is an easy plant. Its foliage suddenly appears in spring and this is soon followed by spikes of large purple flowers. These give the impression of being dusted with a grey bloom. At this stage they are very attractive. By midsummer the foliage has gone, quickly collapsing and dying down, but as it is finely cut and thin in texture it dries out and disappears rather than lying and rotting. It can seed itself among permanently visible ground-covering species, complete its annual growth cycle in this company and leave no harmful trace when it returns to rest.

CYANANTHUS (Campanulaceae) includes some very fine species. A few are at home only on the well-drained slopes of a rock garden, and so cannot really be recommended for the peat garden, yet fortunately one or two find conditions here to their liking and once more we are lucky that this small number includes the most vigorous and decorative. Cyananthus are deep-rooted plants and take unkindly to being disturbed once established. They are long-lived perennials and require only the minimum of attention. If they are planted out from pots while still young they show no signs of being upset. All are herbaceous, dying back to a mass of crowns in winter, some shoots travelling quite a distance underground before appearing above the soil. They are more or less mat-forming and array their leafy stems and flowers in a circle.

C. lobatus blooms in late summer and autumn. At the flowering tips it may reach up to three inches, carrying its large purple-blue flowers singly on terminal and axillary shoots. The tiny leaves are lobed, pale green in colour and punctuate the very hairy stems. The calyces are green, but densely covered with fine bronzy hairs giving them a dark moss-like look. The rotate blooms, about one inch across, immediately remind one of the periwinkle both in shape and flat open appearance.

C. lobatus var. *albus* has pure white flowers and paler leaves. Although known for over thirty years it is still none too plentiful.

C. lobatus var. *insignis* is a form of the species discovered in Tibet by Frank Kingdon Ward in 1924. It tends to be almost completely prostrate and has flowers up to twice the size. In colour they are royal purple.

C. microphyllus differs from the last species by having simple, uncut foliage. It forms a dense flat green pancake of vegetation, the leaves being glaucous beneath. The method of flowering is similar to the last, although the petals are not so full that their edges overlap. They have a star-like effect. A fringe of white hairs in the throat of the tube is very pronounced. An eighteen-inch diameter circle of greyish green foliage tipped with a ring of violet-blue flowers can be very decorative in August.

CYATHODES (Epacridaceae) belongs to a southern hemisphere plant family allied to the heaths and heathers. Unlike those, however, very few are hardy in British gardens, although in sheltered places one or two may be grown successfully. Obviously a plant meant to fruit, but which does not, is fulfilling only part of its function and unfortunately this is the case with Cyathodes. The fruits are rarely seen. Apparently they are brilliant red berries, and although one might argue that the partially shaded areas of the peat garden are not conducive to fruit production, berries still fail to develop on large plants in much more open and sheltered sites.

C. colensoi is indigenous to New Zealand where it is part of the wild heathland. It is a dwarf shrubby plant of six to nine inches, but spreading at least to four times that amount. The wiry stems may start erect, but they soon arch over and elongate. The small hard foliage is closely packed on these stems so that the spaces between are none too obvious. To say the plant is evergreen may not be inaccurate, but when describing the leaves it must be said that their colour is really pinkish-grey. The greyness is more obvious on the younger parts of the plant, suggesting that initially a bloom covers all, and only later after rain has washed the plants a few times does this bloom recede. The pinkiness always remains. Its flowering season is spring, and it is then that the tightly packed heads of small white flowers expand and open.

CYPRIPEDIUM (Orchidaceae) is a very big genus of which only a few are truly hardy. Those that will grow in peaty soil and are mentioned here have creeping underground rhizomes which should be planted in an open mixture, their tips barely covered by soil. Their main fault is that they tend to start into growth early and are therefore vulnerable to late frosts, so if a northern exposure is chosen for them, growth may be later and in consequence escape damage. The fascination of orchids will assure them a favoured site in the garden.

C. calceolus, although recorded in floras as growing in woods on limestone formations, is not averse to peaty, acid soil in gardens. This is the "Lady's Slipper" which sends up flowering and sterile shoots twelve inches or more in height; these carry one, two or three flowers. The large pouch or labellum is pale yellow, while the wings and hood are chocolate-maroon. Three or four sheathing leaves are carried on each stem. A single plant left undisturbed for forty years may produce a clump bearing fifty flowers. Although a British native species, this is really one which is circumpolar in its distribution. As such it has local strains, some of which have been given specific rank, whereas varietal status is considered by some botanists to be more appropriate. (Pl. 9.)

C. parviflorum is really a North American form of the above. As the specific name suggests it has smaller flowers; it also has red spots on the lip of the pouch.

C. pubescens, still within the *C. calceolus* complex, although also North American, is recorded from farther south in that vast country, and the fact that the lateral petals are said to be greenish constitutes the difference between the two.

C. tibeticum is a rarity, virtually unprocurable at present, but a plant not to be forgotten when collecting is resumed in that part of the world. That it does grow out of doors and is completely hardy is manifest in the way in which dedicated growers, without the benefits of replacement, still produce flowering plants at various alpine shows. The green sheathing leaves, each enclosing the base of the one above, are deeply ribbed on the back. The solitary flowers, up to four inches across, have large reddish to dark purple pouches, while the various appendages on a creamy yellow background are heavily netted with reddish reticulate patterns. The cypripediums are not the easiest of plants to manage, but they are among the most desirable.

DEINANTHE (Saxifragaceae) apparently means wondrous or strange flower. This is a very apt description, and in itself should cause the curious gardener to look further. There are only two species, both of them Asiatic, one with white or creamy flowers, the other light lilac. They lack the usual spot character for the saxifrage family, which normally displays a divided style (commonly referred to as pepper and salt). Here the stigma has five lobes and, as one would expect, the ovary has five compartments. Both species grow in moist shaded situations in nature and only under similar conditions do they grow satisfactorily in cultivation. They are completely herbaceous and spread by creeping rhizomes. Both species are summer flowering.

D. *bifida* has globular-shaped flowers which are creamy white. The plant is taller than D. *caerulea*, having twice or three times the vigour, but the most impressive difference is in the large leaves which are deeply bifid. It grows wild in higher woodlands on the Japanese island of Honshu.

D. *caerulea* was introduced by Ernest Wilson around the beginning of this century. He found it in wet places on cliffs in Hupeh. It does not like sunshine or wind, so should receive shelter from both. Initially the leaves are almost pinkish, gradually becoming green as they mature. A cluster of four coarsely serrated leaves forms the base; in well-cultivated plants these can measure ten by six inches, but when poorly grown their dimensions may be no more than a quarter of this. Above are the unusual downward-facing flowers, produced in panicles. The supporting stems are pink. The flowers are spherical while in the bud stage but open widely, and when turned over the lilac-blue petals will be seen to curl round the numerous light violet stamens. These in turn are arranged in a circle. Healthy, well-grown plants are usually twelve inches or thereby in height.

DENTARIA (Cruciferae) is so akin to Cardamine that occasionally they are considered as one genus. Add to this the synonymy in certain species, one being superseded by another, and it is quite obvious that confusion is bound to exist. In this kind of dilemma the best solution appears to be to retain the names under which the plants are most widely known. Those suitable for semi-shaded situations, and considered sufficiently interesting to be included here, like an open soil so that their spreading roots may pass through the surface layer consolidating their advance by forming masses of tuberous tissue. The very scaly rootstock forms a layer less than an inch below the top of the soil. Division in early spring, just as growth starts, is the best means of increasing the size of planting.

D. *digitata* is a very decorative species from the mountains of west and central Europe. It is more than three hundred years since it was first described and even

28 & 29 *Nomocharis* hybrids (page 93).

30   *Orchis maculata* (*Dactylorchis*) (page 133).

31 *Ourisia macrophylla* (page 135).

32 *Ourisia* 'Snowflake' (page 135).

33   *Paeonia obovata* var. *alba* (page 137).

34   *Phyllodoce breweri* (page 47).

35　*Phyllodoce empetriformis* (page 47).

36　*Phyllodoce* × *intermedia* 'Fred Stoker' (page 47).

37 *Primula chionantha* (page 71).

38   *Primula chungensis* (page 72).

39   *Primula helodoxa* (page 74).

40 *Primula involucrata* (page 74).

41   *Primula nutans* (page 76).

42   *Primula rosea* (page 77).

43   *Primula scapigera*
      (page 77).

44   *Primula sinopurpure*
      (page 78).

45   *Primula serratifolia*
     (page 78).

46   *Primula tsariensis*
     (page 79).

47   *Ranunculus amplexicaulis* (page 140).

48   *Ranunculus geraniifolius* (page 141).

49   *Rhododendron campylogynum* (page 51).

50   *Rhododendron forrestii* var. *repens* (page 53).

51 *Rhododendron hanceanum* 'Nanum' (page 54).

52 *Rhododendron impeditum* (page 54).

53   *Rhododendron lepidostylum* (page 55).

54   *Rhododendron lepidotum* (page 55).

55   *Rhododendron trichostomum* (page 60).

56   *Rhododendron yakusimanum* (page 61).

then must have been much thought of as a colourful plant. The spikes of large flowers are rose-lavender and, when in bloom, create a pinkish haze within the shade. The divided leaves uncurl at the same time as the flowers, and when the final height is reached it may be as much as one foot or fifteen inches.

D. *enneaphyllos* grows to eight or ten inches, and while the inflorescence of pale yellow flowers terminates the shoot it also tends to bend over as though it were too heavy to hold upright. It droops below the level of the whorls of divided leaves. This is an attractive characteristic and gives the plant an appearance of shyness. It can be found in the woods of Switzerland and eastwards along the European Alps into Yugoslavia and down into Italy. Spring flowering, it is completely hardy and provides colour for very little effort. *Cardamine enneaphyllos* is a synonym of this species.

D. *pinnata*, with white flowers, could brighten up a shaded corner even more. Here the flowers, although not small, are not as large as those in the previous species, so that spaces separate the flowers and no overlapping takes place. Its foliage, too, is finer while the shoots are not so robust.

DIPHYLLEIA (Berberidaceae) would be placed in the plant family Podophyllaceae if that were universally recognized. Like all herbaceous plants associated with Berberis it is most fascinating, having a leaf and flower balance peculiar to itself. The name is self-explanatory and means what it says—two leaves are produced on each stout stem. The leaves are large, up to a foot and even more across, and deeply cleft, the two halves having edges which are coarsely lobed and toothed. There are two species known, but only one appears to be in cultivation.

D. *cymosa* is North American and grows in woods at high altitudes in Virginia. Its cymose inflorescence carries a number of small white flowers of little decorative value appearing in late spring or early summer. Neither are the small purple fruits, which develop later, of much significance, but in conjunction with the large leaves it becomes a fascinating plant with which only a few other berberids can compare. It will always stand out from other background plants.

EMPETRUM (Empetraceae) is a small genus of a few species, but with an extensive distribution. Widespread throughout the Arctic and north temperate regions, it is also a part of the flora in alpine zones of South America. It is evergreen, heath-like in appearance, and although not colourful in the floral sense introduces a new texture for winter interest. The edges of the leaves are turned under, reducing the amount of stomata exposed to winds.

E. *hermaphroditum* is probably the most northerly member, found only in the colder areas of the north and, when plotted in Britain, confined to the higher hills. As the specific name indicates, the small pinkish flowers are hermaphrodite and can be followed by purplish-black juicy fruits in autumn. The habit of the long branches of this species, which is sometimes considered simply a form of *E*.

nigrum, is arching, so that it is impossible to give a measurement of its height. It forms tussocks.

E. nigrum, the main species, is completely prostrate. The leaves tend to be more needle-like and the stems, which are reddish when young, become darker with age. It forms a solid mat.

E. rubrum is the one indigenous to South America. It is very similar to our native species and superficially differs only by having a woolly covering. Even this can vary, as a form is listed under the varietal name *tomentosum*. This no doubt is given to those plants which are extremely hairy. The flowers occur in the leaf axils just behind the growing tips. The size of the leaves, too, is smaller than those of northern plants. The three species in all constitute an interesting group.

EOMECON (Papaveraceae) translated could mean "Snow Poppy". It is a mono-typic genus, discovered in Kwangsi Province, China, by the Rev. B. C. Henry. Its habitats he records as banks of rivers. This genus, in some respects, is a grown-up *Sanguinaria canadensis*. It is a perennial, although in some areas its hardiness is in question, its over-wintering rootstock being fleshy underground stems. Planted between two background shrubs it could become permanent and a source of interest.

E. chionantha flowers in June. These have four petals and are pure white, the centre of the flower being filled with a mass of yellow stamens. The terminal flowers can measure almost two inches across, while those produced later from lateral buds may be smaller. The twelve- to fifteen-inch stems are purplish, while the basal leaves, which are deeply cordate, have a scalloped type of edge. The pale green leaves are glaucous beneath.

EPIMEDIUM (Berberidaceae), popular with rock gardeners for a very long time, is known to tolerate shade, and this immediately makes it valued as a woodland or peat garden plant. In general they are Old World herbaceous perennials, ornamental in foliage and flower. They bloom in spring, and while the old foliage may be allowed to remain on the clump throughout the winter it is important that it is cut away before growth becomes vigorous in spring. The rootstock consists of a mass of rhizomes and these are best divided in September. Epimediums like a soil rich in humus. Numerous names and synonyms are involved in this genus, but those given preference which follow seem accurate at the time of writing.

E. alpinum is European. It grows wild in wooded slopes in the south-eastern countries. It is a slender, graceful plant, its flowers and leaves being supported on fine wiry stems about a foot high. The leaves are compound, pale green in colour but greatly flushed with reds and browns. The small reddish-purple drooping flowers appear in May.

E. grandiflorum is native to the main islands of Japan and to east China. It has handsome foliage, the leaves being twice ternate and each shield-shaped leaflet is serrated in outline. The "Columbine" type flowers vary in colour from white to violet and some have been given fancy names. 'Rose Queen' describes quite accurately the shade of bloom in this particular variety.

E. perralderianum is a North African species which grows at high elevations in Algeria. Despite this it is quite hardy and has handsome yellow flowers produced in a loose raceme. Three long-stemmed leaflets constitute a leaf and this rises from ground level; the flowering stem is leafless. The toothed foliage is quite hard and as in *E. alpinum*, is flushed with red.

E. pinnatum, from the Caucasus, also has racemes of bright yellow flowers. The leaves, although compound, vary in their number of leaflets. These are bright green and the edges are spiny toothed. Hairs are also disposed along both pedicels and petioles.

E. × *rubrum* is so named because of the bright red inner sepals. Its parentage is uncertain, but it is presumed to be a cross between *E. alpinum* and *E. grandiflorum*. Up to thirty pale yellow flowers may be carried on a fifteen-inch-high stem. The leaves are divided into numerous leaflets. This is a first-rate hybrid.

E. × *versicolor* is another hybrid, this time thought to be between the Caucasian *E. pinnatum* var. *colchicum* and the Japanese *E. grandiflorum*. The yellow flowers are dominant here, although the pale pink sepals are also evident, and in the form offered as 'Sulphureum' the large pale yellow flowers are particularly fine.

E. × *warleyense* is fairly vigorous, producing flowering shoots up to eighteen inches high. The flower combination is red and yellow, and as usual the leaves are compound. Propagation is a simple matter.

E. × *youngianum* is recorded as being a garden hybrid whose parents are two Japanese species, *E. diphyllum* and *E. grandiflorum*. It carries white flowers and grows to nine inches in height. The best clone to be selected from this hybrid is the form 'Niveum'. While the cultivar name simply indicates that the flowers are white, its stature is dwarf, being no more than six inches to the topmost flower. In May a small group can look most attractive in flower.

FOTHERGILLA (Hamamelidaceae) is a deciduous shrubby genus of four species, all of which are available in commerce. They are sometimes known by their common name of "Dwarf Alder", although that in no way flatters them. Two species in particular are grown for their catkin-like creamy flowers which in fact are virtually dense columns of stamens. These are precocious and are plentifully produced on the short stubby side growths. The foliage, too, is most striking in autumn just prior to its being cast. Plants up to eight and ten feet tall are descriptions which may cause dismay, but, when it can be said that it takes twenty years at least for plants to reach five feet in east coast gardens, this once more stresses how little one need consider the ultimate height of some plants. Although slow growing, suckers may develop and these can be detached with roots if extra plants are desired.

F. major is the tallest species. It has large matt-surfaced leaves. They are broad and almost flat topped, and this part is often very coarsely toothed. One easy way to identify this plant is by the colour of its foliage in autumn, which is yellow.

F. monticola has foliage which turns yellow, orange, red and bronze before being shed and so can be named with confidence at leaf-fall. Both species lose their leaves very quickly and this usually occurs immediately after the first frosts. The foliage does not linger on the bushes once the absciss layers have formed. Both are from the eastern North American states. Their value in a peat garden is in both flowers and foliage. (Pl. 12.)

GALANTHUS (Amaryllidaceae) is the botanical name for the "Snowdrop". To write it in that way seems to imply that there is only one species, whereas there are a number native to the eastern Mediterranean countries. Greece and Turkey particularly seem to be rich in them, and the interested grower need only hear of a new form to be off on its trail. Some species require open warm sites and are not really suitable for the peat garden, but there are some which multiply most satisfactorily in this kind of environment. In the majority of cases it is recommended that if bulbs are to be moved this should be done during the dormant season, but this is not so for Galanthus. The most satisfactory time to transplant them is just after flowering, when they are still in full leaf. The soil should be liberally treated and, even although it may be moist, plants which have been disturbed should be watered in.

G. ikariae is found on Icaria, a small Greek island in the Aegean near to the spot where the mythological Icarus fell into the sea after his flight near the sun. This species grows well in partial shade; in fact there is not a vestige of glaucousness on its broad green leaves. These are healthier if sheltered from the sun. Its large, rounded, pure white flowers will brighten up such areas, and if the inner segments are examined the green marking round the sinus will be found to be extensive. *G. ikariae* subsp. *latifolius* may have broader leaves, but the flowers are smaller and so too is the green marking.

G. nivalis, although listed in some British floras, is still doubtfully native, but it has been naturalized for so long that one cannot be certain. It may have been introduced in Roman times. This is an extremely interesting species with a large number of wild variants. The narrow strap-shaped leaves of the typical form are well known; so too are the nodding white flowers and green marked inner petals. The variety 'Lutescens', found wild in Northumberland, has yellow markings; 'Scharlokii' has a divided spathe which resembles bunny's ears; 'Magnet' has long pedicels which allow the flowers to bounce in the least breeze and 'Atkinsii', one of the most vigorous hybrids in cultivation, is a first-rate plant. And, if flowers are required in October, *G. nivalis* subsp. *reginae-olgae* or *G. corcyrensis* will supply these.

GALAX (Diapensiaceae) is monotypic, and all that is desirable in a woodland plant will be found here. It is of American origin, described as indigenous to open woodland in mountainous districts from Virginia to Kentucky. The plant family includes other attractive and difficult genera, but Galax is certainly not one which is capricious.

G. *aphylla* is an evergreen woodland species, bearing glossy green thick-textured cordate leaves which are almost round and up to six inches in diameter. Occasionally they become flushed with red, but this is an additional, unexplained colourful phenomenon which can be expected and should be accepted without concern for the plant. Although producing fewer flowers than one would wish, two or three spikes of small pure white flowers are carried in packed inflorescences arising from the crown in early summer. By dividing old clumps of the thin underground stems in autumn or spring the area planted can be increased.

GENTIANA (Gentianaceae), named in honour of King Gentius of Illyria, who, it is said, was the first to discover the medicinal properties contained in gentian root, is prized for different reasons today. Many have flowers of great beauty. This is a very large genus with species in both hemispheres. These vary greatly from annuals of a few inches to perennials up to six feet in height. Mat-forming kinds are also common and while, basically, the flowers are similar there are individuals with characters peculiar to themselves. By and large, gentians are meadow plants and as such enjoy the association of other genera. This company is something which some plants are less able to forgo than others, and no doubt many a rare and difficult plant is lost because of monoculture. Plant associations in a garden are difficult; there are too many other factors missing to allow much success and names can be conjured up which are now but memories, making one sad at what has been lost to cultivation. Despite all this, when George Forrest collected and sent home seed of G. *sino-ornata* it was said to be his finest introduction. Certainly it and many more have given pleasure for years. Some gentians are better raised from seed, and these include not only the obvious annuals but those with deep searching rootstocks as in G. *lutea*. Most others may be increased from cuttings, the time of taking these varying with the species. If the plants are herbaceous, obviously this should be done as early in the year as possible, in order to allow time for a resting bud to form before the winter. Those that are evergreen can be taken in June or July. Quite a large number, including some of the more desirable kinds, do not require this sophisticated treatment and will quickly grow away from pieces taken from clumps divided in March and replanted in the open. They are a healthy race and remain virtually free from pests and diseases, their only major requirement being enough moisture in the growing season.

G. *acaulis* and its many closely allied associates can be dealt with in one small paragraph. This is not because these Europeans are of little value, but because for reasons known only to themselves they will either flower in one's garden or they will not. If it be the latter, the sensible thing to do is to accept this and grow only those which are not so fussy. An odd clump, moved around various aspects, may eventually settle in an ideal site, but this is far from certain. The flowers of G. *acaulis* are so decorative and redolent of the Alps that their cultivation is at least tried by all who have seen them. To produce a mat of evergreen ovate leaves adorned with large trumpets of blue is the ambition of most alpine gardeners.

G. *andrewsii* is known in America as "Closed Gentian" or "Bottle Gentian" and never opens its flowers. This unusual factor means that in shape the flowers look like bladders. These can be up to one and a half inches long and are arranged in terminal and axillary clusters, the large royal blue blooms being tipped with white. The stems can be a foot high and bear large spear-shaped foliage.

G. *asclepiadea*, the "Willow Gentian", tolerates a certain amount of shade and enjoys moist conditions. It is native to south Europe, where it grows in meadows and woods. In late summer its flowering shoots reach two feet or more and display clusters of bright blue, heavily spotted flowers. The common name gives the key to the shape of the foliage. As a background filler this species is ideal.

G. *farreri* carries a very famous name, first being discovered by Farrer in 1914. The original introduction seems to have been a particularly light greenish-blue flowered form, a colour very different from that seen today. Nevertheless our present plant is most desirable and the thin foliage appears to be still typical of the original. The markings inside the trumpet are beautiful, for they are green, white and variously spotted. Propagation is simple and consists of dividing the clumps in March. Moisture, not shade, provides the cool conditions this species desires.

G. *gelida* is well worth searching for, as it is likened to a yellow-flowered G. *septemfida*. It is native to eastern Europe, the Caucasus and Persia. Stems up to twelve inches high carry the clusters of bell-shaped flowers, which usually open in August and September.

G. *gilvostriata* is at present lost to cultivation, but its ability to produce so many large bells on such a small plant will one day make its reintroduction certain. Frank Kingdon Ward collected seeds in Upper Burma where it grew among rhododendrons, and the last clump known to exist in this country grew for years in a peaty bank. From tiny rosettes small side runners are produced, and it is at the ends of these that the light blue flowers appear. They are embellished by light stripes and dark spots inside the flower tubes.

G. *gracilipes* is an interesting Chinese species which is best left undisturbed once planted. Its long perennial thong-like roots search deeply into the soil, forming a most efficient anchor. It is completely herbaceous, dying back to a cluster of buds formed at soil level. In spring a number of rosettes form, and although not all produce flowers enough of them do to make this plant colourful. The flowering stems, ascending at first, later assume a prostrate position, so that the purplish-blue flowers, borne singly on the branching stems, lie on the soil. For this reason alone it is useful to have a Saxifraga, Silene or some other plant with ground covering foliage over which the stems can rest.

G. *hexaphylla* is unique. One cannot mistake the arrangement of the blue-green leaves in the young rosettes. They are in sixes, hence the name. It was Farrer who introduced this plant into cultivation; in fact he found it in Kansu on the expedition when he first discovered the species that bears his name. Prostrate stems up to four inches long are normal, and these carry solitary blooms at their tips, pale blue in colour but heavily spotted with green dots, appearing in late summer.

G. *kurroo* is unusual in a number of ways, one being that it flowers very late in the season—so late in fact that seed very rarely ripens. Its narrow leaves, glaucous in colour, form a rosette which in many ways resembles that of G. *gracilipes*. Axillary buds develop into flowering shoots and these lie prostrate on the soil, but the stems follow a zigzag pattern ending in a solitary flower, for rarely are two formed. The petals are shaded rich blue, marked in the tube and throat with green spots. Kashmir and north-west Himalaya are given as the natural distribution of this plant. Seed is the best means of increasing stock and certainly, once established, it ought not to be disturbed.

G. *loderi* belongs to Kashmir also. From a cluster of grey-blue rosettes it produces four-inch-long leafy stems carrying large blue flowers with a feathered fringe filling the gaps between the petals. It may be planted on top of a peat block, but it needs light to encourage it to flower well. The ovate foliage is most distinctive.

G. *makinoi* grows in bogs in the high mountains of Honshu. The erect flowering shoots are clothed with slightly glaucous two-inch-long lanceolate leaves. Six or seven upright, tubular, almost stemless flowers, slaty-blue in colour, terminate these shoots. They are heavily spotted. This plant has a very good overall quality much in keeping with a restrained plant collection.

G. *ornata*, although known for more than 150 years, having first been collected in Nepal by Dr Wallich in 1820, is still an extremely rare species in cultivation. The short dumpy flowers with bright blue petals, green-and-white-striped tube and purple spotting, are arranged in a circle at the ends of the four-inch-long shoots. They more or less face upwards. The short leaves are also narrow and clothe the reddish-purple stems. Constant division and replanting are necessary to keep this plant, and to miss out on this for one year can mean the loss of the species. This is particularly so on light sandy soil where the penetration of the root thongs is poor.

G. *pannonica* is summer flowering and has most distinctive foliage. The elliptic leaves are in the region of nine inches long, and several of them form a leafy rosette. Their margins are rough, a number of prominent veins are present and the whole leaf presents a puckered appearance. Usually a number of fifteen-inch-high flowering stems arise from the crown and, on top of these, in the axils of the upper leaves, the clusters of purple flowers are formed.

G. pneumonanthe, the "Marsh Gentian", is a native species with a limited distribution in this country, although throughout Europe and Asia it is widespread. Its common name is a guide to its cultivation, and the reason why plants may not do well in gardens can be due to the better drainage providing drier soil. It is a lovely species with erect stems bearing numerous heavily spotted upturned trumpets in the axils of the upper leaves. When well grown it is a good garden plant and is best raised from seed.

G. pyrenaica has never been termed easy, but a well-grown specimen attracts everyone's interest. Although bearing the name pyrenaica it is found as far east as the Caucasus. Its most distinctive character is the apparent ten-lobed violet blue flower, giving the impression that this is star shaped rather than the usual five-lobed trumpet, and this is brought about by the largeness of the plicae. The tube is still present and is green in the throat. The plants form mats producing numerous flowering and non-flowering shoots usually in the region of three inches long. In nature it grows in wet peaty places but not in shade, and as this can be provided in our peat garden success is surely guaranteed for some growers. Like so many desirable species the biggest problem is in obtaining plants.

G. septemfida is both well known and represented in gardens, and few plants could be considered easier or more floriferous. A hardy tough perennial from eastern Europe and Asia, it quickly establishes itself by sending its thong-like roots deep into the soil. Its herbaceous nature ensures a crop of fresh foliage annually, and this comes from a very large cluster of buds. Usually the flowering stems, well furnished with oval leaves, grow approximately nine inches high, but, as the large heads of flowers are heavy, the erectness suddenly disappears and the stems then radiate from the centre, displaying in August a circle of deep blue trumpets usually well spotted. *G. septemfida* var. *latifolia* is a particularly fine form with large flowers and broad foliage.

G. sino-ornata needs no description; its place in gardens has been assured from the day of its introduction. Found by George Forrest in the moist pastures of north-west Yunnan in 1904, it has since proved to be an easy plant in cultivated soils. Without question its main attribute is the lateness of its flowering season. This is September and October, months when colours are in the main confined to autumn tints and berries. A small plant can be expected to produce a solid clump of large purple-blue green-streaked trumpets every year. Division of the mass of root thongs in March provides an easy method of increase.

G. stragulata creeps across the surface of the soil, rooting as it spreads. It forms an evergreen carpet of leaves while the stems are red in colour, and out of this the red calyces appear, to be quickly filled by the unusually inflated purple-blue corollas. Despite this swelling at the base, the flowers taper towards the end and, when the petal tips curl over, these are severely constricted. The blooms, like the stems, are prostrate and can be almost two inches long. Masses of flowers are not produced, but those that are provide a touch of the unusual.

G. trichotoma has blue flowers, the shade of which is very difficult to describe. One can only enthuse over it and say it is a rich bright blue, paler inside the throat and trumpets. These are upward facing and measure over an inch in length. The inflorescence is a sort of raceme, two or three flowers being produced at each circle of willow-like leaves, so that from a distance the twelve-inch-high erect stems resemble blue spires. This Chinese species is summer flowering and, although it has not been seen for a few years, it is certain to be reintroduced at the first opportunity.

G. *veitchiorum*, introduced at the beginning of the twentieth century and commemorating the firm of Veitch, is one of autumn's finest flowers. The flared part of the bloom is rich royal blue, while within the trumpet yellow lines streak the middle of the petals. Apart from its flower colour it is easily distinguished by its short broader foliage and the pronounced rosettes which persist all winter, the whole being slightly glaucous. Differing forms have been introduced, some by Ludlow and Sherriff being particularly dark flowered, but all are of high merit. This Asiatic species belongs to the same section as G. *farreri* and G. *sino-ornata* and requires similar growing conditions in the garden.

Asiatic hybrids, of which there are many—not all, fortunately, bearing fancy names—have been much sought after by growers of gentians. This wide range of garden plants is of surprisingly limited parentage, the five main progenitors being G. *farreri*, G. *ornata*, G. *hexaphylla*, G. *sino-ornata* and G. *veitchiorum*. These of course are the finest of the Himalayan species and their progeny could hardly be less than award-winning. Among the earliest to be named were G. × *bernardii*, G. × *stevenagensis*, G. × *hexafarreri*, G. × *orva*, G. × *farorna*, G. 'Devonhall', G. 'Glendevon' and G. 'Inverleith' (Pl. IVa). With the desire to increase the number of hybrids, and in order to supply an apparently insatiable market, some of these were increased from seeds. As this provided a swarm of variants there now exists a great deal of confusion with regard to hybrid names, although stocks of some are still considered to be true. If names don't matter, however, a wonderful galaxy of trumpet shapes and colour shades can be had, and if the end product is simply a bunch of lovely gentians then this miscellany will suffice. The enthusiast, however, likes to refer to his plants by name, and it is in his interest that the discerning specialist nurserymen are in business.

GLAUCIDIUM (Ranunculaceae) is placed in this plant family by Ohwi in his *Flora of Japan*. It can also be found in either of two closely allied families, Papaveraceae or Podophyllaceae, depending on the botanist. Wherever it may finally come to rest will add nothing to the charm of this very beautiful Japanese, semi-woodland plant. It is native to the islands of Honshu and Hokkaido and is part of the woodland flora on the high mountain slopes.

G. *palmatum* is the only species, and from a hard tuberous-type rootstock arise many stout shoots, some up to four feet in height. These carry two large palmate leaves just below the large, four-petalled, pinkish-mauve flowers. A boss of stamens fills the centre of the flower and through this the divided style protrudes. Large winged fruits follow in due course. It grows best in a leafy soil in a semi-shaded site, and shelter from the sun helps the petals to retain longer their delicate colouring. (Pl. 14.)

G. *palmatum* 'Album' and var. *leucanthum* are names which apply to the white-flowered forms and these are pure and unblemished.

HACQUETIA (Umbelliferae) contributes much to the upsurge of plant life in early spring, for its naked flowers start to emerge in February. The partly open flowers, which are really heads of flowers, appear to expand immediately the buds become visible. As the season progresses so leaves appear and the flower stalks elongate until finally, by April, the flowers are held on four-inch-high stems. There is only one species. Dondia was for long considered the valid name.

H. *epipactis* is native to central Europe. It produces clusters of yellow flowers, but these are surrounded by a number of green petaloid leaves, or involucre, which gives the impression that this is a large daisy-like flower. The green fleshy leaves are deeply divided and have serrated margins. This umbellifer resents disturbance once established, in fact its roots go down so deeply that some effort is needed to free it from the soil, and even so the thick thong-like roots will have been broken. Where weeding is not done over-zealously, self-sown seedlings are likely to occur and it is from these that fresh plantings should be made. (Pl. IVb.)

× HEUCHERELLA (Saxifragaceae) is itself a hybrid, the result of a cross between two genera, Heuchera and Tiarella. This was brought about in a garden in France many years ago. Since then other growers have repeated the experiment and some of the progeny have been given fancy names, but it is sufficient here to record and report on the cross.

× H. *tiarelloides* is a spreading plant which grows admirably in partial shade. The average height of its flower stems is twelve inches, and these carry panicles of small pink bells. The clump really covers the ground completely as it spreads. Top dressing with leafy soil in spring helps to retain vitality in the centre of the colony, but to ensure that the plants remain vigorous it is necessary to divide them every few years.

HYDRASTIS (Ranunculaceae) is another of those genera met with only in the plant enthusiast's garden. Although two species are said to exist, one in North America and the other in Asia, it is the American plant that will be discussed. The roots are tuberous and at one time, when ground down, were used in medicine. Propagation by seed and division of established plants are two ways of increasing stock.

H. *canadensis* is a woodland species of eastern North America. In spring it develops a stem of twelve inches or thereby on top of which two rough-textured, much-divided leaves expand. One is invariably larger than the other. Beyond the topmost leaf a small stem carries the flower. Actually this consists of three sepals which quickly fall away, leaving behind a cluster of creamy-yellow stamens and a bunch of green stigmas; no petals exist. Later, red berries develop, and these sit upright above the foliage. The black seeds are encased in the white pith within the fruits. (Pl. Va.)

HYLOMECON (Papaveraceae) is a yellow poppy of quality and adds to the number of species that one knows will respond well in the semi-shaded site. If this genus is retained then it is monotypic, but it has been included in Stylophorum, and some recent authorities even consider it belongs to Chelidonium.

H. japonicum has large golden yellow flowers, only one or two forming at each upper leaf axil, but as many divided leaves are borne on the foot-high succulent stems, flowering continues from early to mid summer. The fleshy roots may be divided in early spring as growth commences.

INCARVILLEA (Bignoniaceae) has always fascinated horticulturists and, although there are only a few species involved, it is good that the majority are hardy. They have the most enormous flowers—considered by some to be too large for the accompanying foliage and height of the plants—but no one can deny them a place in the garden. They do not like a shaded area, are long lived and flower well on a flat site. The root system consists of a large tuber which should not be disturbed once established, and if another planting is required young plants should be raised from seed. It may take up to four years for seedlings to reach flowering size.

I. compacta is one of the dwarf species producing stems which rarely reach six inches. These carry the rose-pink flowers two and a half inches long. The foliage is pinnate and broadly oval with a small point at each leaflet tip. June is the month to see these plants in flower.

I. delavayi has long been used in herbaceous borders, where its two-foot spikes of rosy-red petals, added to the white, pink and yellow-streaked trumpets, provide colour early in the season. The larger leaves divide into a great many lanceolate leaflets. These form a base above which the stems tower. A feature not often observed is the long green seed pod, which is frequently spotted with purple dots. This Chinese species has a variant which comes true from seed and is listed as 'Bees' Pink'. The shade is paler.

I. mairei is the correct name at present for *I. grandiflora* var. *brevipes*. Its distinguishing foliage characters are its two or three pairs of basal leaflets followed by a larger terminal one. The leaves are all basal and the flower scape is usually seen to divide a few inches above the foliage. Each stem may carry one or two large reddish-purple trumpets.

I. mairei var. *grandiflora* differs from the above by having unbranched flowering stems or, if divided, this they do just above soil level. Furthermore, one or two pairs of small leaflets occur at the base of the large leaflet. Two forms of this variety were introduced from the wilds of south-east Tibet and Bhutan by Ludlow and Sherriff in 1950. Instead of applying latinized epithets, however, two cultivar names were given. The earlier flowering one with rose-red flowers is known as 'Nyoto Sama' (L.S.E. 15614), while the one which blooms later and has vivid carmine pink almost stemless flowers is 'Frank Ludlow' (L.S.H. 17250). Both are highly ornamental, having glaucous, pinnately divided foliage most of which lies flat on the surface of the soil.

JEFFERSONIA (Berberidaceae), named after President Thomas Jefferson of the United States of America, contains two dwarf herbaceous species, one in the American continent, the other in Asia. Although related they are not difficult to tell apart. A cool shaded situation in leafy soil suits them best, but the position must be well drained. Seeds sown as soon as ripe, or larger plants divided in spring, are ways of procuring additional plants.

J. diphylla is the North American species found in woods on the eastern seaboard. It is locally referred to as "Twinleaf", not only because each crown produces two leaves but also because each of these leaves is divided into two segments by a deep incision which penetrates to the leaf stalk, and the fact that the base of the leaf is cordate. This gives the effect of a pair of slightly glaucous green wings. In May the pure white flowers appear, carried singly on six-inch-high leafless stalks. Each flower comprises six petals in the centre of which is a prominent cluster of yellow stamens.

J. dubia, native to eastern Asia, carries the synonym of J. manchuriensis, a much more enlightening name than the one which is current. It flowers early, in April to be exact, when the cluster of solitary lilac-purple flowers suddenly erupts, the many leafless wiry stems carrying the tiny purple buds very quickly elongating until they are about four or five inches high. The coppery leaves develop a little more slowly, although some may be in evidence at flowering time, until the clump appears to be covered by numerous leafy umbrellas. Both species should be included in plant collectors' lists.

KIRENGESHOMA (Saxifragaceae) grows wild in woodland on some of the Japanese islands, but even there it is reported rare.

It is completely herbaceous and in this country is best planted in a semi-shaded place where it will also receive a degree of shelter from winds. The soft broad foliage will quickly wither and go brown if drying winds and lack of soil moisture are allowed to affect it. The rootstock goes deep and, like most other woodland plants, it likes a soil full of humus. Its one fault is that its flowering season is late, September and October, and as it is very susceptible to frosts an early one can cause damage before the flowering period is half way through.

K. palmata, the sole species, grows to four feet in height. The thin stems are well furnished with palmately divided opposite leaves, the lobes of which may be coarsely toothed. A feature of these wiry stems is that they may be either green, or purple with a glaucous sheen. There is no doubt that the latter form is superior. The much-branched inflorescences carry the large pendent bright yellow flowers, almost two inches in diameter, and produced in such quantities that the stems are unable to hold them erect. Some support in the way of canes to which the branches are individually tied is to be recommended.

LINNAEA (Caprifoliaceae) honours the name of Carl Linnaeus. It grows extensively in the colder parts of the northern hemisphere and is a most worthy plant to carry the name of the famous Swedish botanist. There is only one species and it is referred to affectionately by its common name of "Twin Flower". It requires a cool moist situation, in nature growing in peaty soils in the shelter of pine woods, and is therefore an ideal addition to the peat garden's inhabitants.
L. borealis creeps over the ground. Its long trailing shoots are clothed with tiny, oval, opposite leaves which are well spaced so that the reddish stems show. The twin pink fairy bells, carried on long stems at the ends of short side shoots, appear in June and July. L. borealis var. americana is larger than the European native plant and responds more favourably to garden conditions.

MECONOPSIS (Papaveraceae) for most people is epitomized in the Himalayan "Blue Poppy", M. baileyi, although botanically its correct name is M. betonicifolia. There are over forty species included in this genus, but it is interesting to note that, apart from one, all are native to the countries associated with the high Himalayan plateau. The exception is the European M. cambrica, known locally as the "Welsh Poppy". The plants in this genus have most beautiful flowers, and although basically similar they are produced in species which behave differently, at least in their manner of growth. Some, like the "Blue Poppy", are hardy perennials and flower every year; some are biennial, or almost so, and flower in that cycle; but one other group, this time referred to as monocarpic, flowers once only and then dies, but it may take five or six years to reach flowering size. It is this last group wherein the lovely rosette-forming kinds are to be found, and while young plants can only be had from seeds, the effort of raising them carries with it the assurance of a glorious long-term reward. Many Meconopsis are shade-tolerant and, while in moist areas of the country certain of the larger growing kinds may persist in open sites, the coolness which is associated with thin woodland or a shrub-protected border is appreciated by them. Hence their inclusion here.
M. aculeata is biennial and usually, when in flower, is in the region of eighteen inches in height. This spike carries numerous large wide-open blue flowers, the shade of which may vary, but as the individual plants spread little and send up only one shoot their close planting shows a most impressive range of hues. The shoots, calyces and seed pods are protected by sharp prickles. A significant feature of the broad, blue-green foliage is the manner in which it is lobed.
M. betonicifolia is the most popular species, having been grown in herbaceous

borders for years. Its name is derived from the fact that its leaves, crenately toothed, are not unlike those of the wild "Betony". This most reliable perennial, up to five feet in height, may be raised from seed or by division in autumn or spring. Because plants raised from seed vary, and not all produce the wonderful blue tone so much sought after, obviously plants increased by pieces taken from good colour forms are more satisfactory. Frank Kingdon Ward is given the credit for introducing this species. Spontaneously occurring hybrids appear where species which cross are grown close to one another, so as Meconopsis growers tend to concentrate their collections some most noteworthy garden hybrids have arisen as a result. A particularly fine form with most intensive sky-blue flowers is now offered under the name of M. × sheldonii (Pl. 24.). M. betonicifolia is one parent while the other is reckoned to be M. grandis.

M. chelidonifolia must have shade. Too much sun and just a little wind can cause its leaves to go brown round the edges. This species is best given a place in the background among the rhododendrons where its purplish black stems may elongate to five feet or more. These are thinly adorned with deeply lobed leaves and are terminated by a branched inflorescence of smallish golden yellow flowers. A perennial, the over-wintering crowns are swollen buds and could be mistaken for bulbs. (Pl. VIIIc.)

M. dhwojii may take two to four years to flower, is monocarpic and has very finely divided foliage. This forms an evergreen rosette up to eighteen inches across which, in the flowering year, withers and dies. Many creamy-yellow flowers are borne in a broad pyramid-style flowering stem. During the vegetative phase the leaves are peppered with purple dots.

M. delavayi is sometimes commercially available, although it is really a connoisseur's plant. It perpetuates the name of a French missionary who collected many specimens in the Chinese province of Yunnan. Although its height rarely exceeds six inches, the rich purple blooms are up to two inches across. The leaves are all basal and quite small.

M. discigera can be identified by the prominent disc-like circle at the base of the style. The plant is more or less biennial, although a few members in a batch of seedlings can take more than two years to flower. The flowering spike resembles a three-foot-high tower of blue. The undivided foliage, almost strap-like and less than an inch broad, forms a rosette which dies back to a resting bud if the plant fails to produce a flowering shoot. The following spring, as this bud unfolds, the numerous inward-curling leaves gradually open out so that the centre is like a golden nest. (Pl. 21.)

M. grandis is sometimes confused with the popular M. betonicifolia. It is quite different, however, at least horticulturally, for its leaves tend to be simple and almost without lobes. It produces large blooms on long pedicels usually with only a few stem leaves present. In some forms the flowering season too is much earlier. One listed under the collector's identification number of Sherriff 600 is an especially vigorous, more leafy, large-flowered plant. (Pl. VIIIa.)

M. horridula well deserves this epithet as the spines which protect all its parts, except the petals, are truly horrid. It is a biennial, with narrow leaves, and as its height is usually fifteen inches or less a small group planted in an open space can be a feature. (Pl. IX.)

M. integrifolia carries the most descriptive popular name of "Lampshade Poppy". Although usually only twelve inches or thereby in height, the huge yellow blooms, of which only two or three are usually borne, can each be more than six inches across; moreover the large central boss of stamens adds to its beauty. Although some plants may flower more than once they are not truly perennial. (Pl. VIIIb.) Flowers of the size of this species interest hybridists and selectors and at least two hybrids of worth have been recorded. The one whose progenitor is *M. grandis* is known as *M.×beamishii* and is now very rarely seen, while *M.×sarsonsii*, whose other parent is *M. betonicifolia*, is a much more reliable plant and is often encountered. It is an interesting fact that hybrids of which *M. integrifolia* is one parent always have yellow flowers. The foliage characters favour the other parents.

M. latifolia is a biennial which is difficult to transplant. Seeds should therefore be sown thinly on the site in which they are to flower, or seedlings in pots should be pricked out while they are still very small. During its first year a few broad, dentate leaves merely indicate its position, but in May of the following year, when the spikes of bright sky-blue open blooms are on display, it has few rivals. (Pl. 22.)

M. napaulensis, though hardly classed as one of the finest monocarpic species, is certainly one of the most reliable. Its large rosettes of deeply incised foliage are most decorative, but when the plants bloom, and the spikes can be over six feet high, the colours may be found to vary a great deal. The tendency in most instances is to produce a colony of flowers of mixed shades. These may be white, light blue, pink, wine or purple.

M. quintuplinervia will be grown whether a peat garden exists or not. This running perennial—underground shoots being responsible for the spread of this species—forms mats of simple glaucous leaves in summer. These are herbaceous and if the site chosen agrees with the plant's needs then flowers will follow in due season. Long leafless stems, sometimes measuring eighteen inches, each carry a single, drooping, lilac-coloured flower enclosing creamy stamens. The modest angle adopted by these blooms has resulted in the plant being referred to as the "Harebell Poppy of the Himalaya".

M. regia well deserves this royal term. Monocarpic, this illustrious member has large, creamy yellow, semi-pendulous flowers. They could almost be termed saucer shaped. The blooms on the inflorescence are well spaced but in large enough quantity command respect. While the plant in flower may be considered majestic, its most attractive stage is, indubitably, its years as a non-flowering rosette, for this crown of permanent leaves sometimes three feet or more across is golden in colour (Pl. 23). This is due to the complete covering of fine hairs which give the foliage the texture of suède. Apart from a few shallow

131

serrations, the margins are entire. This woolly covering, unfortunately, is not without hazard as fluctuating winter temperatures can cause rotting if the foliage lies wet. A pane of glass supported over the centre of the crown can give enough protection. In colder gardens where temperatures remain low this precaution is not required.

M. *simplicifolia* requires no further foliage description except perhaps to say that the glaucous leaves, all basal in this species, taper towards the stalk. The flowers may be light blue or of a darker tone and are supported high above the foliage. Although this species is perennial, and has a crown which can on occasion provide offsets, it is prudent to raise young plants from seed whenever possible. (Pl. 25.)

M. *superba*, despite the suggestion that its specific name implies, requires very little written about it here. The reason for this is that it resembles M. *regia* in all respects apart from having flowers which are pure white and foliage which is silvery.

M. *villosa* is a long-lived hairy perennial, but it is also a plant which resents root disturbance. It is better to raise young plants from seed than disturb established clumps. The lobed leaves are green above but intensely glaucous beneath, and this foliage effectively protects the surface roots from the sun in open places. Plants in partial shade are much healthier, more vigorous and their yellow flowers are bigger. One fascinating feature about the flowers which may interest those who cultivate these poppies is the seed capsule; here it is long and pod-like and sheds its seeds by splitting lengthwise rather than through slowly enlarging pores.

MERTENSIA (Boraginaceae) is widely grown in gardens, but the species mostly cultivated are of dwarf, alpine stature and require a great deal of light. These are not considered here, but one occurring in woodland and in wet meadows in eastern North America revels in the kind of conditions a semi-shaded garden provides. Its one fault is that it dies down in early summer, leaving a bare patch of soil, and unfortunately it is a plant which cannot be grown through some other type of ground-covering vegetation.

M. *virginica* appears in March and from then until May the flowers are carried in clusters on twelve-inch-high arching stems and are at first pink as they emerge from the buds, altering to a smooth glaucous blue as the tubes lengthen and expand. The foliage, too, is broad and blue-green without a sign of pubescence. When the shoots and leaves first appear in spring they are purple in colour and only change with the light. It is much appreciated as a wild flower in its natural habitat of Virginia, where it is affectionately called the "Virginian Cowslip" (Pl. 27). A white form is also available, similar to the species in all respects apart from its being less vigorous.

MONESES (Pyrolaceae) contains a solitary species, but it is one that should be

57 *Richea scoparia* (page 142).

58 *Romanzoffia unalaschcensis* (page 143).

59   *Salix lanata* (page 144).

60   *Salix reticulata* (page 144).

61   *Shortia soldanelloides* var. *ilicifolia* (page 147).

62 *Shortia soldanelloides* var. *magna* (page 147).

63 *Shortia uniflora* (page 147).

64   *Streptopus simplex* (page 98).

65   *Trillium cernuum* (page 100).

66 *Trillium erectum* (page 100).

67 *Trillium grandiflorum* (page 100).

68 *Trillium grandiflorum* 'Flore Plenum' (page 100).

69 *Trillium rivale* (page 101).

70 *Trillium sessile* (page 101).

71 *Uvularia grandiflora* (page 102).

tried out where a comprehensive collection of dwarf woodlanders is grown. It occurs in practically all north temperate countries, being native to pine woods among moss and litter. Although its distribution is a wide one it is classed in many cases as rare or is even conserved. It is best raised from seed.

M. *uniflora* is a lovely little plant. The rounded evergreen leaves form a small rosette and from this develops a stem of four inches or thereby, carrying one completely drooping flower. This is creamy white to light pink, and one character which is very prominent is the long projecting stigma. This is the single-flowered "Wintergreen", with scented blooms which appear during early summer.

ORCHIS (Orchidaceae) is a very large group of plants which abound in marshy ground in many localities. They are not restricted to cooler climes nor to one continent but grow in temperate areas in Europe, North Africa, North America and even on some of the islands in the Atlantic. Names have always caused confusion and, while a number of plants may vacillate between species, others are sometimes placed in different genera. In order to consider this group under one heading, and since only a few species are involved, the generic name Orchis is retained, although reference will be made to the current nomenclature as it affects the individuals. They are all tuberous, in fact the nature of these storage organs is often referred to in the keys to their identification. To begin with, the soil should be deep, moist and fertile. On the whole they prefer to remain undisturbed and for years will flower annually in the same position. The division of clumps in autumn after the tubers are fully developed and their transplanting are not too hazardous, however. As some have thick, thong-like roots it is difficult to separate them without causing some damage, but this should be avoided as far as possible.

O. *elata*, now classed as *Dactylorhiza elata*, is one of the most outstanding hardy species cultivated. Although native to southern France, Spain and Algeria it grows vigorously in British gardens. Some stems measure two and a half feet or more while the actual spike of deep purple flowers can be as much as nine inches. The shape of the inflorescence is tapered and spear-like. The leaves are bright green and each one encircles the flowering stem. (Pl. VIIa.) A white form of this has been found in Spain. If and when it becomes available its inclusion will provide a pleasing contrast.

O. *maculata* embraces a number of subspecies which all make worth-while contributions to a peat area. In the main they have spotted foliage, the dark blotches being very pronounced in some forms. Strong plants may grow to twelve inches, the amount of the stem carrying leaves and that carrying flowers always being an acceptable balance. Here the up-to-date generic name is Dactylorchis. (Pl. 30.)

O. *maderensis* is best compared to O. *elata* as this is its nearest ally, and although quite different, at least horticulturally, is often confused with it. Here the shade of purple varies to rosy purple or even lilac. The labella or broad lip of the flower is also wider. In general, although *O. elata* is taller, *O. maderensis* is broader in its other characters. The foliage is longer and broader while the actual flower spike is pyramidal rather than spear-shaped as in *O. elata. Dactylorhiza maderensis*, which is its present correct nomenclatural position, is native to Madeira.

O. mascula, the "Early Purple Orchid", is widespread in Europe and West Asia. Its flowers are of a rich purple-crimson tone and its leaves are often spotted. It is much smaller and neater than the two mentioned in the previous paragraph, but if planted without competition from the more gross-growing kinds it will not go unnoticed.

ORTHILLA (Pyrolaceae) is just another offshoot of the genus Pyrola and until very recently was included in it. This is a solitary species and the main character which determines its position is the one-sided inflorescence.

O. secundiflora, the "Nodding" or "One Sided Wintergreen", is quickly identified by the way in which the eight-inch flowering stem arches over at the top in the manner of Dicentra. The flowers are relatively large, up to a quarter of an inch across. The petals, which are greenish yellow, do not open out but retain a globular shape. It is one of nature's summer flowers and is represented in the floras of Europe, west Asia and North America.

OURISIA (Scrophulariaceae) is confined to the southern hemisphere and species are found in South America, New Zealand and Tasmania. In nature they grow in the open in moist areas by stream sides, but few object to slight shade in cultivation. This is particularly true in dry soils, for this type of location provides coolness and shelter from drying winds. Ourisias are dwarf creeping perennials and spread by means of their short prostrate shoots, rooting as they go, or by partly subterranean stems which pop up a little farther away from the centre of the clump. They are not fastidious as to soil, can be easily divided in spring and are striking when in flower. The genus commemorates Ouris, Governor of the Falkland Islands in the middle of the eighteenth century.

O. caespitosa is one of the tiniest species. It has prostrate shoots which cover the soil as effectively as thyme. They have opposite, silvery-green leaves, barely half an inch long with slightly notched margins. Up to five large white flowers can be produced in May and June on slender four-inch-high stems. *O. caespitosa* var. *gracilis* is even smaller. It forms a close mat of bright green foliage and has relatively large flowers compared with the minute foliage. As in the species the flowers are not entirely white, being relieved by yellow in the throat and trumpet.

O. coccinea is a beautiful plant from the Chilean Andes with a head of pendent scarlet flowers in an inflorescence similar to that of some North American penstemons, but rather sparsely borne. Even the individual flower stalks are the same shade as the flowers. The bright green, broad foliage is lobed and toothed while the main and secondary veins are well defined in pronounced depressions.

O. *macrocarpa* has fairly coarse foliage and should not be placed near less robust plants which could be smothered by it. This foliage is thick and tough, pale green on top and with a crenate margin. Although not appreciated for its foliage, when the large panicles of white flowers appear in June its less attractive traits are forgotten.

O. *macrophylla* resembles closely the previous one and although the name suggests large leaves it would be difficult to say that they were larger generally. However, when in flower the panicle is much less dense. It grows to at least twelve inches high and carries three or four whorls of flowers. (Pl. 31.)

A hybrid Ourisia which appeared spontaneously in a Scottish nurseryman's collection is of such garden value that it well deserved a cultivar name. It is O. 'Snowflake' and assumed to have O. *caespitosa* and O. *macrocarpa* as parents. The shape of the leaves and their size suggest that this could be true, and when this foundling blooms the whole is covered with fairly large glistening white flowers. (Pl. 32.)

PACHISTIMA (Celastraceae) belongs to a family so far not met with in this book. This evergreen genus is of North American origin and is closely allied to Euonymus, with which it could be confused. There are only two species, easily told apart, and at least one of these is grown in pots and exhibited at rock garden shows. Neither is difficult to please, although the one bearing the larger leaves may be damaged by frosts if a mild autumn has prevented thorough ripening of the shoots and foliage.

P. *canbyi* has small narrow leaves less than an inch long which are minutely toothed. They are gathered near the tips of the shoots and from the same region the tiny red flowers, carried on long wiry pedicels, also appear. The plant's approximate height is six inches, but its spreading potential is proved by the suckers that appear, particularly when top dressed.

P. *myrsinites* can reach two feet or more and is at home when growing through or with some other woody, suckering plant. The foliage is finely toothed and ellip-tic in shape. In some respects the general appearance of the plant gives the im-pression that it belongs to Ericaceae (one of the vacciniums, for example). The reddish flowers, borne in the leaf axils, are brighter than in the other species.

PACHYSANDRA (Buxaceae) could hardly be praised for its flowers; they are usually insignificant or, at best, unimpressive, but it does provide interesting ground cover. The term 'ground cover' sounds dull, and *is* dull when it relates to vast areas of soil covered by some low-growing herb other than grass. It reduces plants to their lowest level, that of simply being used—not enjoyed. But these creeping plants form thick mats of foliage in the densest shade, and their role should not be to plant and ignore. In spongy, moist soil their underground stems quickly permeate an area.

P. procumbens, the "Allegheny Spurge", is an evergreen woodland plant of barely six inches. The soft pliable stems are decumbent and tend to root where they come in contact with the soil. The flaccid foliage is clustered at the ends of the shoots, leaving the exposed lower parts of the stems free of vegetation. It is near the base from where these stems radiate that the short spikes of flowers develop. These are greenish or purplish-brown, but the most conspicuous parts are the stamens' white filaments. It is the least vigorous of the pachysandras.

P. stylosa is Chinese and flowers in April and May. It is more often met with under its synonym *P. axillaris*. This species has a cluster of four or five broad, coarsely toothed leaves at the ends of the shoots, but these are so heavy that the stems are inclined to lie on the soil, where they produce roots. The white to pink-tinged flowers are carried in tight spikes of about an inch and occur along zigzagging stems.

P. terminalis, the Japanese species, hardly requires listing as it must be known to all who grow dwarf plants. The closely knit blocks of vegetation which result from the planting of just a few slips are assurance of its ability to flourish in shady corners. This plant displays its inflorescences so that they may be seen; being terminal, they sit above the collar of leaves. The flowers are white. The foliage is ovate, thick yet flexible and slightly toothed towards the apex. A form with white to creamy-yellow edged leaves is known as *P. terminalis* 'Variegata'.

PAEONIA (Ranunculaceae) includes both shrubby and herbaceous plants, but since the shrubby members tend to prefer shelter and so grow quite large, they will not be dealt with here. Furthermore the point may be made that on the whole paeonias prefer a neutral to limy soil whereas the peat garden is obviously acid. Be that as it may, there are several which grow satisfactorily without the presence of lime and as some are so beautiful it is worth including them. Planted between and among the background shrubs they contribute much by their early display of large colourful flowers, but there are one or two which warrant a more prominent site where only close inspection will reveal their full beauty.

P. cambessèdesii is one of the earliest to flower. It is also one of the less vigorous, for, although it is suggested that eighteen-inch stems are the norm, usually they are less than that. The large deep pink petals meet beneath a mass of yellow stamens, and from the centre of these the fiery carpels seem to burst through. The most individual feature of this plant is the foliage, for while the upper surface is grey-green and the veins are traced in red the whole of the underside is reddish-purple. It is reputed to be a rare plant on the island of Majorca from where it was first introduced; its name honours Jacques Cambessèdes, a French botanist who published a flora of the Balearics.

P. emodi is included because it is white if for no other reason. It belongs to

Kashmir and a lovely plant it is, with that shade of pale green foliage so often associated with white flowers. It also carries two or three flowers on each shoot, a feature of the group comprising the Lactiflora section. Like most species it is better raised from seeds as old plants lifted and divided may sulk for years.

P. *mlokosewitschii* varies in colour, but in the better shades of yellow it is out-standing. The huge petals curling to form a cup, as though protecting the inner floral parts, contrast well with the red shading on pedicels and leaf stalks. The compound leaves (which occur in all paeonias) are produced in sufficient quantity to form an effective background. As the stems are unable to remain upright without help at flowering time some form of support should be given. This should be done early in the season and before the flowers are showing. Birch twigs are effective, although wire mesh can be just as serviceable.

P. *obovata*, like all beautiful things, fades too quickly. If only one species could be planted this, or to be more precise the white form of it, var. *alba*, would be the one chosen (Pl. 33). Apparently only the white form is in cultivation, a situa-tion which must be unique among plants, for usually it is the most sought-after member which cannot be had. P. *obovata* must be given a forward position. The glaucous bloom which covers the shoots and broad foliar segments in itself places this species above the general run of green-leaved plants. But when the alabaster white flowers open, revealing the yellow stamens, the red to lilac fila-ments and the prominent, eye-catching crimson stigma, a feeling almost akin to worship affects the beholder.

P. *tenuifolia* is different from all other paeonias in that the leaf segments are so much cut and reduced in width that they bear more resemblance to a fern than a flowering plant. It is always less vigorous, usually attaining less than fifteen inches in height, and the flowering stems produce solitary deep crimson flowers. It is eastern European and has been in cultivation for more than two centuries.

P. *veitchii*, collected in Szechuan in 1907 by Ernest Wilson, has never been either rare or out of cultivation since. It is a very long-lived plant with pale green leaves, the leaflets themselves being deeply incised. It does not have the largest flowers in the genus, but two or three are carried on each shoot. These are light magenta and show a slight degree of fluting at the edges. Self-sown seedlings are common. The variety *woodwardii* is slightly less robust and usually paler.

PLATYCODON (Campanulaceae) is the Japanese "Balloon Flower" and was given this name on account of its inflated flower buds. These balloon out and remain closed until the flowers are almost fully developed. Their position in the peat garden is on an open terrace and, as the rootstock is a thickened one, obviously root disturbance should be discouraged. There is only one species, widely scattered through the mountains of the Japanese islands, and also occurring in China, Korea and Manchuria.

P. *grandiflorum* is so variable in height and flower colour that cultivated names are sometimes applied to segregated forms. In the main it has large Campanula-like bells, some more open than others, and the colour may range from purple-blue to pure white and sometimes pink. The usual height of the plants is in the region of twelve inches, and so some of these have been used as herbaceous border perennials for years, but in other instances six inches or less may be a more accurate statement and plants of that stature require a little more careful cultivation. In recent years *P. grandiflorum* 'Apoyama' made its debut. This form carries enormous violet-coloured blooms, three inches across, on stems just over six inches and, as in the other varieties, on top of the usual broad, slightly serrated, blue-green foliage and glaucous stems.

POLYGALA (Polygalaceae), the "Milkwort", although a very large genus of a few hundred species, contains only a small number hardy in colder zones, and of these two are considered here. They are sub-shrubs, evergreen and increase by underground suckering stems. Plants placed a few inches apart in open soil soon close the gaps and form a dense mass of shoots. To propagate these plants cuttings may be taken in summer which will root in a cold frame and be established by late autumn, or rooted pieces may be taken from the clumps and planted straight out in spring.

P. chamaebuxus, the "Box-leaved Milkwort", has a distribution which embraces most of the countries in central Europe. It is found in a great many of the highest alpine woodlands running through the other vegetation. It flowers all summer and the blooms have erect, white, rounded petal-like structures. The lower keel is yellow or orange tipped, so that the flowers are quite obvious among the linear, evergreen foliage. *P. chamaebuxus* 'Purpurea' is a selected form in which the wings of the flowers are rosy-purple.

P. vayredae is closely allied to the last species and is probably a western form. It grows in Catalonia, in the eastern Pyrenees. Every visible part of this plant is thinner than in *P. chamaebuxus*. This slimness gives it a very wiry look, but in pattern of growth it follows the other species. In *P. vayredae* the flowers are always magenta-purple while the crest at the end of the keel is yellow tipped.

POLYGONUM (Polygonaceae) includes some pernicious weeds, a fact which has affected its image, but outwith the many less attractive species there are several that are dwarf, colourful and not too rampant. Some will form a blanket-like growth over the ground, rooting into the soil where the nodes are in contact with it; others, less inclined to leave stem rooting to chance, have subterranean shoots which root as they pass through the soil. None is difficult, although one or two could be a trifle tender. The general distribution of Polygonum is world-wide, particularly within the temperate zones, and from such coverage one can expect great variations.

P. affine has been cultivated for many years, and the *Botanical Magazine* Plate 6472 shows the familiar thin-spiked, narrow-leaved plant. In recent times plant hunters have introduced forms from different localities and these have been found to vary considerably from the original. This is to be expected, so obviously the better forms are the ones that should be planted in gardens and those which follow are in that category. This species forms mats of vegetation which change in autumn and become very bright and colourful before withering. The leaves vary from narrow to broad and taper towards the base. The flower spike, too, may be long and thin or short and stout and carry both pink and rose-red flowers, the shade depending on the age of the individual blooms. 'Darjeeling Red' and 'Donald Lowndes' are two distinct forms which should be chosen.

P. *emodi* is likened by slme to P. *affine*, but it is vastly different. Its stems are much narrower and zigzag over the ground, travelling farther in a season than do the other species. The leaves, too, are not so broad, are sharply pointed and thicker and harder in texture. No flower variation is obvious on the narrow flower spikes and the flowers themselves are red. It is also less hardy, but as it roots very readily young plants should be over-wintered in cold frames in case of emergency.

P. *tenuicaule* is a most delightful miniature of six inches or less. Its stems are stoloniferous and broadly ovate, radical leaves up to two inches long develop where the stems break the surface. The small two-inch-high inflorescences carry tiny white flowers, but as they are produced in number they are quite conspicuous. This Japanese native species can take over a peat block quickly.

P. *vacciniifolium*, at home in the Himalayas, will find many of its high plant associates in the peat garden. In many ways it is similar to P. *affine*, flowering, colouring and dying in the same seasons and manner. Here, however, the spikes are much narrower, almost sharp pointed, and the foliage is ovate. It is worth noting that although the leaves are green the undersides display light glaucous colouring. The flowering period from August to November is both colourful and long.

PTERIDOPHYLLUM (Papaveraceae) is yet another Japanese monotypic genus scarcely in cultivation at present, but one which should be carefully tended and increased where possible. The difficulty is that it belongs to the Poppy family and that group is notoriously unco-operative when it comes to dividing or transplanting. The name is most descriptive and means 'fern leaf'.

P. *racemosum* grows in coniferous woods on Honshu and is reported to be rare even there. The pinnate leaves are approximately six inches long and while, in a general sense, the plant is deciduous, sometimes the foliage can persist throughout the winter in sheltered corners. Its flowering season is during the summer months and it is then that the nine-inch-high spikes of tiny white poppy-like flowers appear.

PYROLA (Pyrolaceae) is one of nature's aristocrats. Its presence in the garden indicates not only an interested plantsman but also points to a successful cultivator. The qualities of daintiness and charm, added to that of rarity, combine in making one aware of a choice species. Difficult? Yes! Impossible? No! And one of the most suitable places for this gem is underneath the north side of a medium-sized Rhododendron. Careful top dressing rather than the use of hand-fork and trowel is best since it spreads by underground stems.

P. media is summer flowering, normally during August. The leafy part of the shoots at ground level form a base from which the twelve-inch leafless flowering stems arise. Seven or eight light pink flowers with rounded petals forming a hood are well spaced on these shoots.

P. minor is the "Lesser or Small Wintergreen". It also has pink flowers which in the bud stage are very like pink orbs. They are attached alternately to the upper part of a stout nine-inch stem. In the more shaded sites its natural spread is often more rapid.

P. rotundifolia, "Round-Leaved Wintergreen", is well described both in its scientific and popular names. Creeping through pine needles and other vegetation, this species sends up foot-high stems on which large, open, pure white flowers are borne. The elongated, leafless spikes add gracefulness to its summer display.

RANUNCULUS (Ranunculaceae) is the botanical name for a great many well-loved plants of which "Goldilocks", "Spearwort", "Buttercup" and "Crow-foot" are some. These are admired and acclaimed when seen in the open meadow, but very few could be termed garden plants. Certain species in fact are considered troublesome weeds. There must be few genera, however, that cannot support at least one highly bred lady of which it can be proud, and it is from those termed 'aristocratic' that the gardener makes his selection.

R. aconitifolius is sometimes grown in beds surrounding a pond or even in lighter areas in woodland settings, but it really presents no challenge to the cultivator. It grows too tall, up to four feet, and its white flowers, which in fairness could not be called unattractive, are rather small. The leaves are very similar to those of Aconitum, hence its specific name. The garden form with double flowers, *R. aconitifolius* 'Flore Pleno', widely known as "Fair Maids of France", is well worth growing. It is usually only about two feet in height, which commends it to the gardener with limited space. The more solid blooms intensify the light colouring and bring a better balance to flowers and foliage.

R. amplexicaulis, a Pyrenees mountain native, has been cultivated in gardens for many years. It flowers in April and May and during that time its brilliant white petals, surrounding the cluster of yellow stamens, are much admired. Yet its interest extends beyond the flowering period. Its stem-clasping, glaucous green, broad but slender pointed foliage makes it attractive for as long as it remains above the soil. The branching stems are usually six to nine inches high, each carrying a few flowers. (Pl. 47.) Even in nature the species has good and better forms, so it is not surprising that gardeners seek the one listed as *R. amplexicaulis* 'Grandiflorus'. Self-sown seedlings of this form are not uncommon and are true to size.

R. geraniifolius, perhaps still better known as *R. montanus*, the "Mountain Buttercup" or "Molten Gold", is a garden-worthy plant of low stature, no more than six inches, and very floriferous. In May it completely hides its much indented bright green foliage with large, circular, single flowers. It spreads slowly by underground tubers and these should be divided from time to time in the same way as other herbaceous plants when the centres of the clumps show signs of weakness. (Pl. 48.)

R. gramineus varies a great deal, some strains being tall and spindly while others are short and squat. The narrow glaucous leaves are in themselves attractive and the plant is further enhanced by large, bright yellow flowers which appear in spring. It is found in south-west Europe. Although it is perennial, occasional seed sowing keeps the stock vigorous. A rare and most delightful hybrid known as *R.* × *arendsii*, presumably raised by the German nursery firm of Arends and reputed to be a hybrid between *R. amplexicaulis* and *R. gramineus*, is a charmer. It grows no taller than six inches, carries two or three reasonably large, rich creamy-coloured flowers on each branching stem and has fairly narrow more or less basal leaves which tend to be a shade of yellow-green.

R. lyallii is the large white "New Zealand Buttercup" with the upstanding umbrella type foliage. Sometimes it flowers quite well; at other times it sulks and grows very little. The peat garden may not be the ideal habitat, but in drier areas it is possibly the only place where sufficient soil moisture is available. It is certainly worth trying and, if the reward is a branching inflorescence of large, white, semi-double flowers, the plant variety in the peat garden will be greatly enhanced. Even the twelve-inch-wide circular and saucer-shaped foliage is not unattractive.

R. parnassifolius is just one other example which proves that nature's wild species have a quality that man's induced hybrids cannot achieve. This graceful mountain species, indigenous to the alpine regions of southern Europe from north Italy westwards to Spain, has thick, dark green, heart-shaped leaves, although in some instances the cordate base tends to be less pronounced. The leaves are mostly basal apart from the cluster of leafy bracts which congregate where the flowering stalk divides into several branches. The pure white flowers, pink-flushed on the outside, can measure an inch across and this on a plant of only six inches! An open moist situation suits it best, where it will flower in early summer.

R. thora should be included for interest. It has a tuberous type of root which makes it a good perennial and from which the six- to eight-inch-high flowering stems arise. The flowers are small and sparse, but the glaucous-green foliage, thick-textured and with sessile stem leaves of an indeterminate shape, has crenate margins.

RANZANIA (Berberidaceae) honours Ono Ranzan, a celebrated Japanese naturalist, who has even been referred to as the 'Linnaeus of Japan'. This plant adds yet one more Japanese monotypic genus to the collection and is said to be a rare endemic of Honshu. One would be wrong to class it as a colourful floriferous species, but it would be equally wrong to ignore its out of the ordinary appearance. It is a woodland plant in nature, and while alpine house treatment is meted out by plant enthusiasts in this country this is hardly necessary, for it is hardy and enjoys the peat garden.

R. *japonica*, sometimes placed in the plant family Podophyllaceae, has a rhizomatous root which travels slowly through the soil. In the same way as do other herbaceous members of Berberidaceae it suddenly erupts through the soil and, without first establishing a base of vegetation, quickly runs to a height of twelve inches. Two soft and coarsely lobed ternate leaves then begin to develop, but, at the same time and from between their point of union, out pops a single pendulous six-petalled flower, bright lavender pink in colour, and suspended on a long pedicel. Partial shade ensures that the pale flower colour does not fade quickly and also provides the coolness which is essential for healthy growth.

RICHEA (Epacridaceae) is a southern hemisphere shrubby genus, the main centre being Tasmania, and in a general way, apart from a few sheltered gardens in the south-west peninsula, was not considered hardy in Great Britain. This last statement has been proved incorrect on more than one occasion, for even on Scotland's east coast, at the Royal Botanic Garden, Edinburgh, three species grow out of doors and two of them have lasted for forty years. When judged by their foliage they do not resemble Ericaceae, the family to which they are closely allied, but suggest that their true affinity might be with Dracaena, Cunninghamia or even some of the bromeliads. The tough, sharp-pointed, evergreen leaves are very distinctive, for they clasp the stem with their enclosing bases, and apart from flowers add a new leaf pattern to the area.

R. *scoparia* will grow to five or six feet and is a much-branched plant, the stems of which are densely clothed with two- to three-inch-long pointed leaves. These are evergreen but often show a bronze tint. The terminal spikes, up to four inches long and carrying numerous flowers, can be white, white tinged pink, or orange, all three forms being represented in the peat garden at Edinburgh. It flowers during May and June. (Pl. 57.)

ROMANZOFFIA (Hydrophyllaceae) is a small spring-flowering genus confined to western North America and eastern Asia. The species are not difficult to grow provided there is enough moisture present in the atmosphere to keep the leaves healthy. They dislike a light sandy soil, and if this is the nature of the loam the addition of peat will help to counteract drought. In a superficial way they can be likened to the white-flowered *Saxifraga granulata*, even to the bulbous nature of the resting buds.

R. *californica* has white petals and a tube which has a yellow throat. The flowers are carried in loose, branching, leafy inflorescences up to six inches high. The larger, lower foliage is rounded with a deep cordate base, while the leaf margins are prominently crenate.

R. *unalaschcensis* is the more compact species in that its basal foliage is more dense and overlapping, the leaves are lobed rather than crenate and the flower spike is not so elongated. While still leafy, the flowering stems seem tighter and more distinct and carry the white flowers closer together. This is the better plant. (Pl. 58.)

SALIX (Salicaceae) is a very large genus widespread throughout the northern hemisphere, in particular the temperate and subarctic zones. They are deciduous trees or shrubs and of the latter some may be so dwarf that the height could almost be ignored. In the main they are precocious in their method of flowering and usually the flowers are borne in prominent, erect catkins. Much use has been made of them in gardens, and in the context of the peat garden some of the dwarfer species can find a home. They enjoy moist conditions and the fact that they are deciduous can be of value, one reason being that while shelter from the sun may benefit certain neighbouring plants in summer, winter light is so poor that plants which cast their leaves help to counteract this. Depending on the species, some may prefer a place in the solid peat and colonize a peat block, while others, bearing long graceful branches and planted on top of a peat bank, will be able to display their form more effectively. While this may not always be significant, if the plants are to be grown for their catkins then obviously it is the plant carrying the larger male flowers which should be chosen.

S. *apoda* is an eastern European dwarf shrub of a few inches. It spreads quite quickly and to help it remain low the branches should be pegged down. This will allow the plant to display its catkins better. These appear in March and are at first silvery white with wool, but very quickly the yellow stamens turn them into bobbins of gold.

S. *arbuscula* is British and forms a twiggy shrub up to two feet. Its old branches are brown in colour, but are a bright green for much of their first season. The catkins are not highly decorative, but the glabrous foliage is blue-green on the under-side.

S. × *boydii* is one of those chance hybrids sometimes introduced into cultivation from a wild locality, but which, having combined a unique set of characters, are no longer found in a wild state. This plant is well established in gardens and slowly develops into an erect branching bush. The few dowdy grey catkins produced are of little consequence, but the blue foliage which is assumed to be influenced by S. *lapponum* is infinitely decorative. S. *reticulata* is presumed to be the other parent.

S. *hastata* 'Wehrhahnii' is also much in demand and must be one of the most attractive flowering dwarf willows in cultivation. From strong suckering shoots,

143

side growths appear and on these the catkins develop. In the first instance they are more like round balls of white wool, but by the end of April they have increased lengthwise and are suddenly cream coloured and spiky under the influence of the stamens.

S. herbacea, the "Least Willow", is a tiny charmer which slowly forms mats of thin twiggy growths. The terminal catkins are small and are not readily observed as the plant is in leaf by the time they appear. The small leaves are almost orbicular and have a crenate margin. It is not a difficult plant to increase, as a single small specimen dealt with as one would a herbaceous plant in March will provide many small rooted pieces which will establish quickly in a moist soil.

S. lanata can be too spreading if left unpruned, but as it does not take unkindly to secateurs there is no reason for excluding it on that score. It has beautiful, broad, woolly foliage, silver-white in fact, and while the leaves are still young it is as bright as many plants in full flower. Large catkins develop just as the leaf buds start to break; although the female plant in this instance is as colourful as many other male willows, it still lacks the impact of the yellow pollen provided by the male form. (Pl. 59.)

S. myrsinites forms a shrub up to eighteen inches with loosely interwoven branches. When in flower these display conspicuous purple anthers. The foliage, too, is of interest as it is distinctly shiny green and the margins are roughly serrated.

S. repens may be a little aggressive in a small garden, but can be confined to an allotted area by dint of judicious pruning. Where there is space its long prostrate shoots form an attractive pattern, and in April, when the flowers open, the upstanding catkins look like stubby candles arrayed along the branches. They are quite bright while the anthers are shedding pollen. The under sides of the small, narrow, green leaves are covered with soft silky hairs. There is a very striking variation, *S. repens* ssp. *argentea*, where this silkiness also occurs on the upper sides, giving the plant a silvery appearance.

S. reticulata must rate as number one among the dwarf species. It has so much to commend it that the danger lies in over-emphasizing its virtues. The "Reticulate Willow" gets its name from the prominent veins which criss-cross the grey-green leaves, being distinctly glaucous underneath. They are deeply recessed on the upper side of the broadly oval, inch-long foliage. The stems are completely prostrate and form a close woody network from which the terminal and lateral shoots develop. The catkins do not appear until May, being carried on young leafy shoots, but what does make a most interesting association is the planting of both sexes together so that after a few seasons both the creamy yellow male and purplish female catkins are seen, interspersed throughout the mat of foliage. (Pl. 60.)

S. retusa, the "Blunt-Leaved Willow", is so prostrate in growth that no height can be given for it. As the name suggests, the leaf apex is blunt. The leaves are normally bright green and shiny and little more than half an inch long, and in

May and June the tiny greenish catkins appear among them. This is a European native, but its distribution does not include the British Isles.

S. serpyllifolia has a similar distribution to that of the previous species and has quite often been considered a small form of it. Apart from size, which is about half or even less, it looks the same, with polished light green foliage and a mat-like growth. It gets its name from the diminutive dimensions of its leaves, which are similar in size to those of *Thymus serpyllum*, the common thyme.

SANGUINARIA (Papaveraceae) or "Bloodroot", this referring to the red sap which exudes when the rhizome is wounded, is a dwarf eastern North American genus of a single species. It grows in woodland and by mountain streams, which would indicate that the plant does prefer moist conditions. If left undisturbed it will spread slowly, but if it has to be moved the best time for carrying out this operation is in August just after the leaves finally die. The thick rhizomes should be damaged as little as possible.

S. canadensis, being herbaceous, rests all winter, but towards the end of March the soil above the thick rootstock starts to bulge, and very soon afterwards what looks like growths of greyish-purple mushroom heads are seen coming through the soil. A few days later they have developed into the white flowers of this species. Usually there are eight celandine-like petals, but the flower stalk is hidden by the wrapped-round, rugose leaf blade. This finally opens out to show a large kidney-shaped leaf with a deeply scalloped margin supported on a stout red stalk. The flowers of the single form are rather fleeting, as is normal in the Poppy family, but there is a very full, double-flowered form, *S. canadensis* 'Flore Pleno', which is superior by far, the flowers being carried on four-inch-high stems. (Pl. XIIIb.)

SAXIFRAGA (Saxifragaceae) includes a very wide range of plants. Of species, forms and varieties there are many, and of these probably a lot would grow in the peat garden. Some of course would be quite unsuitable, but the few considered seem to fit in with the general plant association. The open rock garden species tend to have hard, reduced foliage and prefer a really sharply drained soil, but from the others, which frequent moister sites in nature, an interesting selection can be made.

S. brunoniana requires a moist site near the front of the bed. From a single green rosette planted in spring a large number of red, thread-like tentacles radiate. These can travel twelve inches and at their tips small buds are carried, and establish new colonies where they come to rest. Above this strawberry-like bed arise the red-stemmed, branching inflorescences, consisting of numerous small bright yellow flowers.

S. cuneifolia belongs to the same section as does "London Pride". It is a woodlander and as such is worth considering for places where light is restricted. It forms small evergreen rosettes which gradually spread, rooting as they go, but while they may form weed-suppressing ground cover, it is only after they have been replanted into fresh soil that practically every rosette flowers. These are small and white.

S. *fortunei* is a highly decorative plant and is probably within the ken of every gardener. This plant is herbaceous and certainly hardy in so far as its roots and resting crowns are concerned. The top growth quickly succumbs to freezing temperatures and collapses like a Begonia immediately following the first sharp frosts. This is a little unfortunate as the spectacular, tall inflorescences appear only in October, but if planted within the protection of tall shrubs this may counteract the first onrush of winter and allow a little more time for flowering. The white spidery flowers bring lightness to a dark corner. Apart from the flowers this species is grown for its foliage—which is almost round but deeply lobed. In addition to this there are colour foliage forms where the leaves show no green at all but are pink below and reddish-purple on top. Cool conditions are necessary if this soft, succulent foliage is not to be spoiled by the edges drying out and turning brown.

S. *granulata*, the "Meadow" or "Bulbous Saxifrage", can sometimes be found in the turf beneath the cool shade of trees, but is not really a plant for shade. It deserves an open site low down in the peat garden where an ample supply of moisture is assured. Small bulbils constitute the rootstock and the dividing of these makes propagation simple. The plant usually grows in clumps and forms clusters of small leaves with rounded lobes. Those on the flowering stems are reduced to a few fingers. Its white flowers appear in April and May and the whole plant withers and dies by early summer. A double flowered form, 'Flore Pleno', is more usually seen in cultivation.

S. *pensylvanica* blooms in summer. It has twelve-inch-long lanceolate basal leaves which are pale green and have serrated margins. From this lush growth the three-feet-high flowering stems emerge. These are sometimes branched and carry numerous small greenish-yellow flowers. This is more of a curiosity than a riot of colour.

SHORTIA (Diapensiaceae) is among the élite of garden plants. Its foliage and in particular its flowers, have qualities sought for and rarely found. Only nature could produce such balance and design. In recent literature modern botanists have increased by one the number of species included in this genus. They have now placed a closely related plant, Schizocodon, within its framework. In many ways this seems a pity as the name Schizocodon is affectionately known to many. Moist, partially shaded, peaty sites are ideal for members of this genus, as in nature, both in eastern North America and Japan, it is mostly in alpine woodland that they are to be found. In gardens they prefer a sheltered site protected from the hottest sun and certainly from searing winds, but not necessarily under the canopy of a taller plant. Propagation has always been a problem and, while they can be divided, the plants show suckering; the re-establishing of the pieces is both slow and tricky. If the site is moist, shaded and remains so, and the soil into which the small portions have been planted is fine and contains some sand, it can be accomplished out of doors. Ideally it is better to pot up these divisions and place the pots in a north frame where the atmosphere is controlled. Planting out from pots presents no problems. These may be placed in the vertical sides of the north-facing peat banks where the darkness of the medium and the shade of the site form the ideal background to the light-coloured flowers. Seed, although sometimes available, takes years to reach flowering size and, as the small seedlings remain long in a vulnerable state, losses can be high.

S. galacifolia is the North American species. The evergreen ground cover is formed by tough rounded leaves, scalloped or toothed at their edges, which grow from the short creeping stems. Sometimes the petioles are long, giving the leaf a spoonlike appearance. In colder weather extra colouring is added to the leaves in the form of red and crimson tones. The long flower stalks are often red and each bears a single nodding, white to pink, open bell-shaped flower. The edges of the petals are beautifully frilled and undulating.

S. soldanelloides (*Schizocodon soldanelloides*) is easily identified from the other species by the acute likeness of the flowers to Soldanella. Moreover the flowers are borne in a one-sided raceme which seems to add to the crimped appearance of the blooms. Up to ten flowers may be counted on a single stem. These are tubular at the base, but the broad tips of the petals are laciniate and can best be likened to a deeply cut fringe. The colour is deep to light pink, the denser shades being confined to the tube. The leaves of this species are not so thick as those of the other two; they also vary in size and shape, which has resulted in latinized varietal names being given. *S. soldanelloides* forma *alpina* is one of the smaller, more compact forms, the leaves having a cordate base. The difference seems to be solely in the degree of daintiness. *S. soldanelloides* var. *ilicifolia* (Pl. 61) is well named as here the cuneate foliage resembles the holly in shape if not in size or colour, while in the variety *magna* (Pls. XIIIa and 62), which seems to be the one most plentiful in nature, the large almost round leaves, generously toothed and cordate at base, can be more than three inches across. Pink and magenta shading can occur on the older leaves of this species while the younger ones may be brownish-orange in the developing stages.

S. uniflora is an exquisite dwarf Japanese sub-shrub which grows in high woodland on Honshu. One cannot but be affected by its beauty. The slender creeping stems congregate at their ends a few, prominently toothed, evergreen leaves. It is not uncommon to see some plants display reddish shades on the margins. Beautiful, large, delicate pink flowers are produced singly on the four- to six-inch-high scapes. They measure up to one and a half inches across and are wide open with a short tube, the petals, once more, being irregularly frilled and the yellow stamens conspicuous. (Pl. 63.) The larger-flowered forms are segregated under the varietal name of *grandiflora* in cultivation, some of which have been given cultivar names.

SORBUS (Rosaceae) are found in all north temperate continents and are, of course, mostly trees of some stature. There is at least one species of dwarf habit which can be included in the planting and has something to contribute.

S. *reducta* grows to only two feet at the most and this does not take too long to attain. It produces seed like its taller associates, but in addition spreads by suckering shoots. These are not slow to become established and within a couple of years are flowering, in typical rowan fashion, later providing autumn tints with pinkish fruits and coloured pinnate foliage. It was discovered only in 1943 in Hupeh. Although there appears to be more than one clone, these differing only slightly such as in their degree of suckering, it is in a sense a good species and as such comes true from seed.

STYLOPHORUM (Papaveraceae) consists of three species of which one is in general cultivation. In some respects it resembles the "Welsh Poppy", a lovely plant if it were not such a prodigious seeder. Luckily Stylophorum does not have this failing. It does produce seeds, but these are not troublesome and any increase, unless it be on a large scale, can be had by dividing the crowns in spring.

S. *diphyllum* is the "Celandine Poppy". It favours a partially shaded site where its stems will reach up to fifteen inches. These are topped by a pair of divided leaves which are deeply lobed, from between which one or more large, four-petalled golden-yellow flowers occur in May. A bluish tint covers the leaves. This species is the only one in eastern North America, where it is native to woodland areas.

SYNTHYRIS (Scrophulariaceae) is a dwarf perennial herb which looks like a small Veronica when in flower. It produces short racemes of blue or violet flowers in April, and as some species are herbaceous the flowers are developed before the basal leaves have been formed. The rootstock is a short stout rhizome which sometimes stands proud of the soil and should be top dressed, but if extra plants are wanted it is quite a simple matter to remove a few pieces of rhizome in early May, just after flowering, and replant them.

S. *missurica* produces many small purple-blue flowers on a six-inch raceme. Singly these do not constitute a display, but once a crown has formed, giving rise to a number of flowering stems at the same time, it is quite effective. The basal leaves are round and cordate and have doubly toothed margins. It is found growing at altitudes of over 7,000 feet.

S. *reniformis* has round foliage not unlike the last species except that here the leaves are more or less hairy while in S. *missurica* they are glabrous. This is the coastal species occurring below the 3,000-foot mark.

S. *schizantha* is one of the taller plants producing a flower spike of over eight inches. It really forms a herbaceous-like clump and, while the basal leaves are hardly present when the first flowers unfold, they are certainly well developed before the flowers are over. The deeply cordate leaves are densely toothed, deep green in colour and are between two and three inches wide.

S. stellata has a short raceme of lilac blue flowers. They appear early in the year when, apart from a number of smaller leaves on the flowering stem, no foliage is present. It is shorter than the others and on average rarely exceeds six inches in height. This is an ideal species for including in the front row of plants in the peat garden.

TANAKAEA (Saxifragaceae) is rarely met with in gardens today, although at one time it was much more often seen. It is another monotypic Japanese species, but even in an up-to-date flora of that country it is reported as rare in the wild. It is interesting to note that the plant is dioecious, that is to say there are male and female plants, and the female does not produce runners or stolons while the male plant does. The creamy-white flowers are small and produced in long slender compound racemes up to ten inches high. They resemble dwarf astilbes. Yoshio Tanaka (1838–1916) was an eminent Japanese botanist, after whom the genus was named.

T. radicans, the specific name describing the stoloniferous nature of the stems, is an evergreen suited to a moist shady corner. The elliptical leaves, slightly pointed and toothed at the apex, are long stalked and have a thickened succulent-like or leathery texture.

THALICTRUM (Ranunculaceae), given the vernacular name of "Meadow Rue", is a large genus with practically a worldwide distribution. It is a reliable decorative perennial, some species being popular as herbaceous border plants, and is admired for its attractive much divided foliage as well as its light, graceful inflorescence. The peat garden, however, has not too many spaces for tall herbaceous plants, but in the gaps between the shrubs room can surely be found for at least one of the more decorative species. They enjoy rich, well-drained soil, but do not object to a little light shade; in fact their flower colour seems more pronounced where the plants are sheltered from the sun. The word 'tepals' is sometimes used where the perianth parts, sepals and petals, are very similar and difficult to tell apart. This is the case with Thalictrum.

T. chelidonii is a glabrous perennial up to seven feet, but most of this height is composed of the narrow inflorescence, a multiple panicle which casts no shadow whatsoever. It carries the numerous pendent flowers, which owe their colour to the yellow stamens and large mauve petal-like sepals. It is a most beautiful plant but requires the support of a stake.

T. diffusiflorum is a most distinctive species, having grey-blue foliage which is divided into very small segments. It too can grow up to seven feet, but usually its height is little more than half of that. In this species the tepals are pale lilac and approximately three-quarters of an inch long, with the suggestion of a slight bloom on the outsides. They are carried on long pedicels which gives their drooping habit a graceful form.

T. reniforme is the third Himalayan species. It is closely allied to *T. chelidonii* but is glandular. The flowers are also larger, being up to one and a half inches wide. They are pinkish-purple and are pendulous. Like the other two species, young plants should be raised from seed, and it is only in a cool moist atmosphere that the best results are attained.

TROCHOCARPA (Epacridaceae) is very closely akin to Pentachondra and Leucopogon. Of the six species of which it is composed, three are endemic to Tasmania. Like so many dwarf antipodean evergreens, one of the main problems in growing them is to provide sufficient shelter from the cold drying winds of early spring which damage their foliage. A sheltered nook in the rock garden or a place in a protected site in the peat garden make the cultivation of this plant possible out of doors.

T. thymifolia flowers in January and February (midsummer) on Mount Wellington, Tasmania, and in this part of the world they open in June. This dwarf shrub, which may take many years to reach twelve inches, produces a dense mass of twiggy branches which completely mask the soil. If it has one fault it is that the green of the tiny foliage tends to be too dark. The long petiole is recurved and this brings the leaf blade into an apparently more protected position. Flower buds form in early spring but take a long time to develop. They are carried in short terminal nodding spikes and are red in colour.

VANCOUVERIA (Berberidaceae) commemorates Captain George Vancouver, a British explorer who carried out surveys off the western coast of North America. It is a herb found in deep shade in that part of the world. "Redwoods" and "Douglas Fir" form some of the overhead canopy. They are all good reliable perennials for part or full shade. Allied to Epimedium and easily grown and propagated, they spread slowly through the loose mould by creeping rhizomes. Two species have leaves which are persistent while in the other they are deciduous. May is their flowering month.

V. chrysantha has biternate leaves which form a ground cover up to eight inches high, and above this tower the loose panicles carrying the small yet attractive yellow flowers.

V. hexandra is the deciduous species. Here the leaves may be tri-ternate, so dividing the foliage into much finer segments. The flowers are white and in the loose inflorescence look very much like tiny stars.

V. planipetala, in its native haunts referred to under its vernacular name of "Redwood Ivy", has hard tough dark glossy green leaves which seem to be present for ever. They are compound, the leaflets being further divided into five segments. The many-flowered panicle (sometimes as many as fifty may be displayed), carries its white flowers on thin wiry pedicels.

SELECTION OF PLANTS
FOR THE SMALLER PEAT GARDEN

It is sometimes quite difficult when confronted with a long list of names to choose plants most suited to a particular type of situation. As a guide, which is by no means comprehensive, I have compiled lists of plants which will provide colour or interest, or both, and yet be less than eighteen inches in height. I trust these will help the reader round his dilemma.

SHRUBS

(Some of these may eventually exceed the ideal height when fully grown, but their contribution while small is so valuable that, initially, they must be found a place.)

Arcterica nana
Cassiope lycopodioides 'Major'
,,   'Edinburgh'
Cyathodes colensoi
Enkianthus perulatus
Epigaea gaultherioides
Gaultheria cuneata
,,   procumbens
,,   trichophylla
Harrimanella stelleriana
Kalmia angustifolia 'Pumila'
Kalmiopsis leachiana 'Marcel le Piniec'
Pachistima canbyi
Pernettya tasmanica
Phyllodoce aleutica
,,   × intermedia 'Fred Stoker'
,,   nipponica
× Phyllothamnus erectus
Rhododendron calostrotum
,,   campylogynum
,,   ciliatum

Rhododendron fastigiatum
,,   forrestii var. repens
,,   hanceanum 'Nanum'
,,   impeditum
,,   imperator
,,   lepidostylum
,,   microleucum
,,   nitens
,,   pemakoense
,,   radicans
,,   sargentianum
,,   scintillans
,,   williamsianum
,,   yakusimanum
Salix apoda (male form)
,,   reticulata
Sorbus reducta
Tsusiophyllum tanakae
Vaccinium caespitosum
,,   delavayi
,,   nummularia

HERBACEOUS and EVERGREEN PLANTS

(These are suitable for planting on the level terraces and open slopes.)

Adonis vernalis
Astilbe simplicifolia
Calanthe alpina
Calceolaria tenella
Cardamine asarifolia
Chionographis japonica
Clintonia andrewsiana
Codonopsis dicentrifolia
,,   ovata
Corydalis cashmiriana
,,   cheilanthifolia
,,   nobilis
Cyananthus lobatus var. insignis
Cypripedium calceolus
Deinanthe caerulea
Disporum smithii
Epimediun × youngianum 'Niveum'
Erythronium californicum
,,   revolutum
Fritillaria meleagris
,,   pallidiflora
Galax aphylla
Gentiana gracilipes
,,   hexaphylla
,,   pneumonanthe
,,   septemfida
,,   sino-ornata
,,   veitchiorum
Hacquetia epipactis
Heloniopsis breviscapa
Incarvillea compacta
,,   mairei var. grandiflora
Lilium oxypetalum
Meconopsis integrifolia
,,   quintuplinervia
Nomocharis mairei
,,   saluenensis
Omphalogramma elegans

Omphalogramma vinciflorum
Orchis elata
,,   maderensis
Ourisia macrophylla
Paeonia tenuifolia
Platycodon grandiflorum
Polygonum hookeri
Primula alpicola
,,   aureata
,,   chionantha
,,   chumbiensis
,,   cockburniana
,,   edgeworthii
,,   gracilipes
,,   involucrata
,,   luteola
,,   melanops
,,   nutans
,,   obtusifolia
,,   polyneura
,,   reticulata
,,   scapigera
,,   serratifolia
,,   viali
,,   whitei
Pteridophyllum racemosum
Ranunculus amplexicaulis
,,   gramineus
,,   parnassifolius
Sanguinaria canadensis 'Flore Pleno'
Saxifraga fortunei
,,   granulata 'Flore Pleno'
Streptopus roseus
Synthyris missurica
,,   reniformis
Tanakaea radicans
Tricyrtis hirta 'Alba'
Trillium chloropetalum

Trillium erectum
,, grandiflorum
Uvularia grandiflora

Vancouveria chrysantha
,, planipetala

PLANTS for inserting in the face or on top of the peat walls.

Andromeda polifolia
Arcterica nana
Bruckenthalia spiculifolia
Calceolaria tenella
Cassiope 'Edinburgh'
,, fastigiata
,, mertensiana var. gracilis
Chiogenes hispidula
Cornus canadensis
Cyananthus lobatus
Disporum smithii
Empetrum nigrum
Galax aphylla
Gaultheria cuneata
,, depressa
,, nummularioides
,, procumbens
,, thymifolia
,, trichophylla
Harrimanella stelleriana
Kalmiopsis leachiana 'Marcel le Piniec'
Linnaea borealis
Ourisia coccinea
,, macrocarpa
Pachistima canbyi
Pachysandra stylosa
Pernettya tasmanica
Philesia magellanica
Phyllodoce aleutica
,, caerulea
,, empetriformis
,, × intermedia 'Fred Stoker'
,, nipponica
Polygala chamaebuxus

Primula aureata
,, boothii
,, calderiana
,, edgeworthii
,, geraniifolia
,, gracilipes
,, whitei
Pyrola rotundifolia
Ranzania japonica
Rhododendron campylogynum
,, camtschaticum
,, cephalanthum var.
crebreflorum
,, forrestii var. repens
,, impeditum
,, imperator
,, myrtilloides
,, pemakoense
,, prostratum
,, pumilum
Romanzoffia unalaschcensis
Salix herbacea
,, repens
,, reticulata
,, serpyllifolia
Shortia soldanelloides
,, uniflora
Tanakaea radicans
Trientalis europeus
Trochocarpa thymifolia
Vaccinium caespitosum
,, macrocarpon
,, nummularia
,, vitis-idaea

SHADE-TOLERANT SPECIES

(A number of those included here are extremely desirable plants.)

Anemonopsis macrophylla
Arisarum proboscideum
Arum italicum
,,　maculatum
Asarum europaeum
Cardamine asarifolia
Deinanthe bifida
,,　caerulea
Dentaria digitata
,,　enneaphyllos
,,　pinnata
Epigaea asiatica
,,　gaultherioides
,,　repens
Galax aphylla
Gaultheria adenothrix
,,　miqueliana

Gaultheria nummularioides
,,　procumbens
,,　yunnanensis
Leucothoë axillaris
Liriope muscari
Pachysandra procumbens
,,　stylosa
,,　terminalis
Paris polyphylla
Polygonum japonicum
Ruscus aculeatus
,,　hypophyllum
Saxifraga fortunei
Tanakaea radicans
Vancouveria chrysantha
,,　hexandra
,,　planipetala

INDEX